MW01620203

Hansjörg Lipowsky
and Emin Arpaci

Copper
in the Automotive Industry

1807–2007 Knowledge for Generations

Each generation has its unique needs and aspirations. When Charles Wiley first opened his small printing shop in lower Manhattan in 1807, it was a generation of boundless potential searching for an identity. And we were there, helping to define a new American literary tradition. Over half a century later, in the midst of the Second Industrial Revolution, it was a generation focused on building the future. Once again, we were there, supplying the critical scientific, technical, and engineering knowledge that helped frame the world. Throughout the 20th Century, and into the new millennium, nations began to reach out beyond their own borders and a new international community was born. Wiley was there, expanding its operations around the world to enable a global exchange of ideas, opinions, and know-how.

For 200 years, Wiley has been an integral part of each generation's journey, enabling the flow of information and understanding necessary to meet their needs and fulfill their aspirations. Today, bold new technologies are changing the way we live and learn. Wiley will be there, providing you the must-have knowledge you need to imagine new worlds, new possibilities, and new opportunities.

Generations come and go, but you can always count on Wiley to provide you the knowledge you need, when and where you need it!

William J. Pesce
President and Chief Executive Officer

Peter Booth Wiley
Chairman of the Board

Hansjörg Lipowsky and Emin Arpaci

Copper in the Automotive Industry

WILEY-VCH Verlag GmbH & Co. KGaA

The Authors
Dipl.-Ing. (TU) Hansjörg Lipowsky
Werkstofftechnische Beratung
Blaschkestr. 8
85120 Hepberg

Dr. Emin Arpaci
Ingenieurbüro Dr. Arpaci
Büro für Werkstoffe
Naumannstr. 81 H. 38
10829 Berlin

In cooperation with
ECI – European Copper Institute
Avenue de Tervueren 168, b10
1150 Brussels
Belgium

Library of Congress Card No.: applied for

British Library Cataloguing-in-Publication Data
A catalogue record for this book is available from the British Library.

Bibliographic information published by the Deutsche Nationalbibliothek
The Deutsche Nationalbibliothek lists this publication in the Deutsche Nationalbibliografie; detailed bibliographic data are available in the Internet at http://dnb.d-nb.de.

Typesetting K+V Fotosatz GmbH, Beerfelden
Printing betz-druck GmbH, Darmstadt
Bookbinding Litges & Dopf Buchbinderei GmbH, Heppenheim
Cover Design Adam-Design, Bernd Adam, Weinheim

Printed in the Federal Republic of Germany
Printed on acid-free paper

ISBN 978-3-527-31769-1

Contents

Copper in the Automotive Industry. Hansjörg Lipowsky and Emin Arpaci

ISBN: 978-3-527-31769-1

Preface

Copper, gold and tin were the first metals which mankind learnt to use. Copper and gold are amongst a few metals that can be found in nature as solid metals. As copper can be worked easily the oldest known civilisations used it to make useful items as early as 10,000 years ago.

In spite of this people nowadays are less aware of copper than of other industrial metals. Many engineers in the automotive industry are surprised that copper is even used in automobiles. At best mechanical engineers think of copper in connection with the board circuit and starter and generator.

Indeed wire harness, generator, auxiliary motors and electric actuators make up the lion's share of the copper in a car. The implementation of hybrid drives and fuel cells will also see a further increase in the use of copper.

The hybrid drive saves fuel and leads to less harmful exhaust gas. The change from hydraulic to electric actuators and power functions benefits the car and the environment. Electric functions are better suited to consume energy only when there is an actual demand, can be better controlled intelligently, and operating liquids like hydraulic and brake fluids do not need to be disposed of at the end of the vehicle life.

Copper also serves for dissipating waste heat of the control units of power electronics or from engines. The above mentioned applications use copper as it's electrical or heat conductivity is higher when compared to other technical metals.

Most automotive and mechanical engineers are not aware of the many advantages which copper alloys can have for mechanical parts. Of great benefit is the precise hot and cold formability which enables the production of precision parts which, because of their shape, cannot be produced by machining.

The excellent machinability with long tool life and high precision often results in parts for instance in brass being less expensive than in other metals which may be considerably cheaper. Well known gliding properties of copper alloys and the high wear resistance of some special alloys can be combined with very satisfactory strength properties.

Copper is environmentally friendly. It is essential to man, even indispensable for maintenance of the body's defences. It is germ-killing, which may be highly desired in air conditioning units and water pipes. Copper and its alloys are corrosion resistant in various environments (see section "Corrosion"), as centuries

Copper in the Automotive Industry. Hansjörg Lipowsky and Emin Arpaci

ISBN: 978-3-527-31769-1

old copper roofs and Bronze Age artefacts show, which have lain for millennia in the soil. Copper is perfectly recyclable. Secondary copper has absolutely identical properties as primary copper. Therefore no particular specifications (see Annex) are needed for remelted copper and alloys.

The mission of this book is to spread this information primarily to the automotive industry, but also to other fields, and to support them with data and facts.

Hepberg, July 31, 2006 *Hansjörg Lipowsky*

1
Raw Material Resources

1.1
Primary Raw Materials [1]

The earth has an average copper content of approximately 0.006% [1]. Copper is the twenty-third most common element in the earth's crust [2]. There are traces of copper in almost all types of rock. Like iron, copper has the tendency to combine easily with sulfur and oxygen, which accounts for why both metals are often found together in ore as sulfurous minerals.

Copper is seldom found in metallic form, although there are examples in the Urals, at Lake Superior in the USA and in New Mexico. The most important copper ores are copper pyrite (chalcopyrite) and copper glance (chalcosite). Copper pyrite ($CuFeS_2$) contains 34% copper and copper glance (Cu_2S) 79%. Important ores in mining include sulfide mineral peacock ore (bornite) and the oxide minerals: malachite, blue malachite (azurite) and red copper ore (cuprite).

1.2
Availability

Copper ores are mined in underground and open pit mining. Figure 1 shows the main copper mining areas: in Africa, e.g. Zambia and the south, the west coast of South America, central and northern Chile, Peru, Mexico, the lakes area of North America, Canada, the south west of the USA and the former Soviet States: Russia, Kazakhstan and Uzbekistan. There are also significant deposits in Australia, China, Indonesia, Papua New Guinea and the Philippines. In Europe, the only deposits worth mentioning are in Portugal, Poland (Upper Silesia), Serbia and Bulgaria. Chile with 35.5% of mining output worldwide was the largest copper producer in 2005.

Known copper reserves are estimated at 940 million tonnes (2004), of which 470 million tonnes are commercially extractable under current conditions. Potentially usable reserves of copper are currently estimated at 1.6×10^9 tonnes. There are more reserves in maritime "manganese nodules", which are today not

1) Unless stated otherwise, these details refer to per cent by weight.

Copper in the Automotive Industry. Hansjörg Lipowsky and Emin Arpaci

ISBN: 978-3-527-31769-1

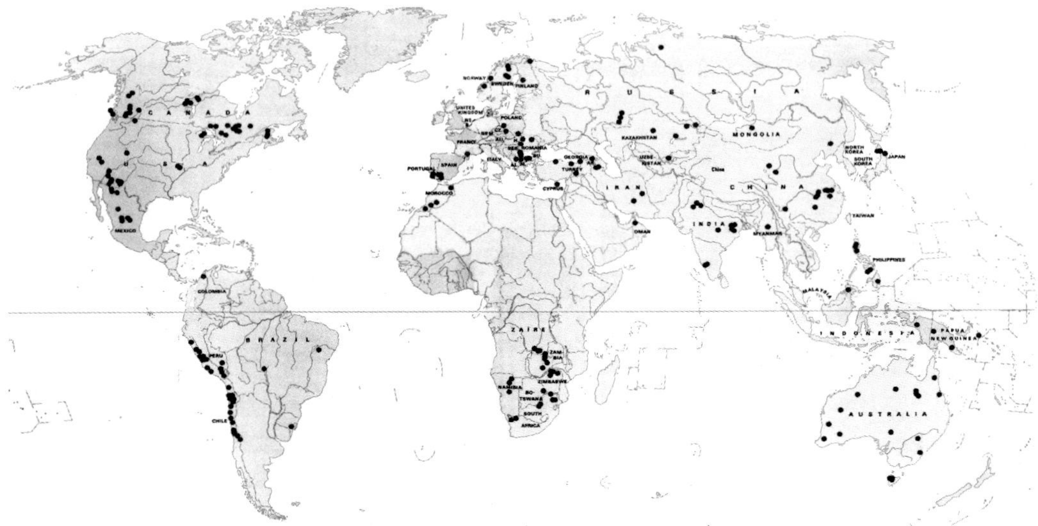

Fig. 1 Copper in the World [2]: The mining output of copper ore in 2003 produced around 14.4 million tonnes of copper [3 a]

commercially viable. The copper content of the manganese nodules alone is estimated at 0.7×10^9 tonnes [3]. Work is now being carried out on processes for commercial copper extraction from these deposits, which are low in metal or difficult to mine.

The supply of copper in the form of deposits and reserves has constantly increased over the years and is thus assured for the foreseeable future.

2
Production

2.1
From Ore to Copper Concentrate [1, 4]

Crude copper ores have a significantly lower copper content than pure copper minerals. The ores mined nowadays often contain only about 1% Cu and even as little as approximately 0.35% Cu in some larger mines. The latter can only be economically extracted in the more cost-efficient open pit mining using state-of-the-art extraction methods. The terrace-shaped open pit copper mines are the largest ore mines in the world. Although much less copper than iron is produced, the amount of rock displaced corresponds to that of the world's entire iron ore mining.

Large quantities of "barren" rock (gangue) are separated from the copper-bearing ore before smelting. The ore is crushed and pulverised into particle sizes often less than 100 μm. Sulfide copper ores are enriched into concentrates by flotation whilst the minerals are separated from one another by various surface properties. The concentrates normally have a copper content of between 20 and 30%; 50% in very favorable cases.

In contrast, copper is extracted from mixed sulfide oxide or oxide ores (about 15 to 20% of all copper ores) by combined special processes or by hydrometallurgy whereby the copper is dissolved from the crushed ore with acid and precipitated by electrowinning.

2.2
From Copper Concentrate to Refined Copper [1, 4]

Copper concentrates are nearly always processed pyrometallurgically. Formerly the process involved three processing steps: partial roasting followed by smelting into copper matte with a copper content of 30 to 50%, which is then processed in the converter into blister copper (crude copper with a copper content of 96 to 99%) and fire-refined into copper anodes with a copper content of ≥99% and oxygen content of <0.2%.

Today, the flash smelting process (e.g. Outokumpu process) is generally used [5], which is especially cost-effective for large quantities of material. The pre

Copper in the Automotive Industry. Hansjörg Lipowsky and Emin Arpaci

ISBN: 978-3-527-31769-1

dried concentrates are simultaneously roasted and smelted in a reaction shaft using oxygen-enriched blast air at 1300 °C, with a hearth area underneath to separate the resulting matte and slag. A waste heat boiler and filter have been added to the exhaust gas line to cool down the gases and separate flue dust from the gases. Sulfuric acid is produced in contact towers from the filtered furnace gases containing SO_2.

The copper matte is periodically tapped out of the furnace hearth and transferred to the converters. The remaining iron sulfide is oxidised by blowing air into the copper matte whereby sulfur is discharged as SO_2 from the converters with the waste gas and subsequently used for the production of sulfuric acid. The iron is oxidised into iron oxide, forming a slag in combination with added silica. The remaining copper sulfide is finally decomposed, the sulfur being extracted again as SO_2.

Direct processes which unite all the reaction steps – roasting, smelting and converting – in one process have had considerable success in recent times. The Mitsubishi and Kennecott processes enjoy large-scale use in this area.

Crude copper extracted by pyrometallurgical means is firstly refined in a molten state (fire refining) and then by electrolysis. Fire refining without subsequent electrolysis is only performed to a very small extent nowadays.

In fire-refining (solidified crude copper or scrap, see secondary copper production), impurities can be removed by injecting air into the copper melt in the anode furnace. The remaining sulfur is thus removed and later the oxygen content reduced in rotary furnaces (anode furnaces) by injecting a reducing gas like methane or propane, natural gas, naphtha or ammonia. In former times, and still today in areas where gases are not available or expensive, the reduction step can be performed by immersing wooden logs into the liquid copper, thus creating a strongly reducing atmosphere. The fire-refined tough-pitch copper with a copper content of $\geq$99% and an oxygen content of $\leq$0.2% is partly cast in continuous casting plants into shapes, such as billets, slabs and ingots. However, the larger part is cast into anodes on casting wheels and subsequently refined by electrolysis.

Impurities in the copper have to be eliminated or reduced to a minimum of some ppm (parts per million), as even very small amounts can seriously affect copper's thermal and electrical conductivity. In the electrolytic tankhouse, the anode sheets cast from fire-refined copper and thin cathode starter sheets made of electrolytic copper, or long-life stainless steel cathodes in the case of more modern plants, are suspended in tankhouse cells filled with a copper sulfate solution. The anode copper dissolves on applying an electric voltage and is deposited on the cathodes as very pure copper while the impurities and accompanying metals either dissolve in the electrolyte or sink to the bottom of the cells as anode slimes. The copper cathodes thus produced are sold or remelted and cast into shapes (billets, cakes and wire rod) of semi-finished products.

2.3 Secondary Copper Production

About 5.4 million tonnes of refined copper were used in the EU-25 in 2004, of which 3.379 million tonnes were produced in Europe. The quantity produced from secondary materials totalled about 2.269 million tonnes (42%) [3]. The share of secondary copper used every year from the recycling of scrap and materials extracted from the overall copper production process is 43%.

Materials have been recycled to extract copper for thousands of years [8–10]. For a long time, the valuable metals were collected in recycling containers and returned to the respective metal cycle after correct preparation and smelting. This is why most materials containing copper are recycled after usage – recycling is not a modern invention! Copper can be repeatedly smelted, refined if necessary, and processed into new products without loss of quality. The quality of copper extracted in secondary metal production processes is equal to that from primary metal production processes with no limitations. For this reason, no distinction is made on the metal markets as to whether the copper was produced from primary or secondary materials. Furthermore, substantially less energy is used in the production of refined copper from secondary materials than for primary copper production.

2.4 Energy Consumption [12]

The consumption of primary energy for fabricating semi-finished products depends on the nature of the semi-finished products, e.g. strips, tubes, etc., and on the material from which they are to be manufactured, i.e. copper or a copper alloy. Thirdly, it depends on the nature of the raw material, i.e. whether the copper was extracted from an ore with a high or low copper content and whether the ore was mined in open pit or underground mining.

The energy requirement in the production of copper semi-finished products (mining, smelting and subsequent processing) is approx. 50–60 GJ/t (13.9–16.7 kWh/kg). Energy consumption can however be reduced to 35–45 GJ/t (9.7–12.5 kWh/kg) if the repeated recycling of the copper is taken into account. Spreading the initial high energy consumption over subsequent recycling processes involving less energy consumption is justified as copper does not undergo any loss of quality when used again and again.

The requirement for primary energy is 30–40 GJ/t (8.3–11.3 kWh/kg) of semi-finished product when the energy consumption for mining is excluded.

Primary energy consumption is calculated at 20–21 GJ/t (5.6–5.8 kWh/kg) of copper cathodes when only the refining of copper is taken into account.

Primary energy consumption for smelting copper cathodes, casting and producing profiles, tubes, strips, etc., is 8–13 GJ/t (2.2–3.6 kWh/kg) of semi-finished product. Energy consumption is determined by the technology used and the quantity of foundry returns. Foundry returns and clean scrap can normally be reused directly without further refinement and therefore with no corresponding primary energy requirement.

Primary energy consumption in the production of copper wire rod from cathodes is determined by the wire size and insulation. Consumption is 8–18 GJ/t (2.2–3.6 kWh/kg) of wire rod (without the consumption for insulation). Considerable savings (over 60%) can be made through modern processing using one heat (e.g. casting and rolling into wire rod).

Continuous and semi-continuous smelting and casting processes with decisive advantages regarding energy consumption, yield and emissions are largely used for the production of semi-finished products in copper and copper alloys.

Modern smelting technology has decisive advantages regarding energy consumption and environmental friendliness. Energy savings of over 30% can be achieved.

The study [12] gives more details on energy consumption for certain semi-finished products and copper alloys. Energy consumption for a number of semi-finished products can be modularly composed and calculated for the various copper alloys as the results of the study are in the form of individual modules.

2.5 Recycling [8–10]

Recycling saves resources, reduces environmental pollution and saves energy. Copper is one of the most commonly recycled materials in the world. Acquisition and trade with scrap and salvage of copper and copper alloys has been well organised for generations due to copper's high economic value and excellent intrinsic properties for recycling.

The achieved product quality, besides the energy saving, is crucial to the assessment of the production of metals from recycled materials and scrap. If this quality is not achieved, the advantages of energy savings would be cast into doubt as the energy consumption of dissimilar materials is compared. In contrast to copper, it is, for example, simply not possible with some other metals and synthetics to manufacture products of the same quality as they are when new. With copper on the other hand, products can be manufactured without any quality loss and such products differ in no way from those produced from primary metals. This applies even after copper has been recycled numerous times, irrespective of whether materials low or rich in copper with metallic or non-metallic parts are added in the recycling.

The "classical recycling rate" is calculated from the annual output produced from secondary materials in relation to the total annual output. Copper has a recycling rate of approximately 43% according to this definition, but this number says little about the actual recycling of material, as it does not consider that salvage is derived from durable industrial goods produced at a time when annual copper production was considerably lower. In calculating the traditional re-

Table 1 Service life (in years) of various copper products.

Small electric motors	10–12
Motor vehicles	15–18 (vehicle service life)
Cables	30–40
Buildings	60–80

cycling rate, the secondary copper output arising from the earlier low production is related to the very much higher current production. This classic recycling rate is misleading inasmuch as it does not express the true degree to which formerly produced copper material is actually recycled.

A "genuine recycling rate" for copper is achieved if it is based on the service life of the products until recycling and the amount of recycled copper in relation to overall production output at the beginning of its service life. Copper and copper alloys have a long service life due to their excellent durability as Table 1 shows.

If, according to this, an average service life of 35 years is taken as a basis, a "genuine recycling rate" of nearly 80% results, based on the overall copper production 35 years ago, and will rise further in future due to the development of suitable separation processes for complicated electronic scrap. This rate describes the almost complete recycling of secondary copper better than the classic recycling rate. Rather, the latter expresses that approximately 43% of overall copper demand is covered by scrap and recycled materials. Material returns are an important source for meeting copper demand as e.g. for Germany which has practically no copper deposits of its own but a well-working remelting industry.

About 14 million cars are taken off the road every year in the EC. The majority of these are recycled or scrapped. The recycling rate of parts of scrapped cars, by European Directive, will reach 85% by 2006 and 95% by 2015 [13].

When considering the quantities of recycling materials in copper production, there is a distinct difference between direct scrapping in works producing semi-finished goods and the use of material returns for the production of refined copper as not all scrap has to be processed by a recycling smelter into refined copper. Works producing semi-finished goods take back sorted production scrap from their customers, such as chippings, chad and compact scrap of the same composition, and remelt it to produce semi-finished goods. For example, rod made of leaded free-machining brass is produced from almost 100% recycled material. Also the ingots produced for casting in foundries where recasting is carried out (mainly cast brass and red brass) are nearly always produced only from scrap and material returns.

There are the following main groups when producing secondary copper taking the *origin* of the material returns into account:

- *New scrap*: production waste from the production of metal and semi-finished products and sorted chippings and chad without major contamination

- *Old scrap*: return of end-of-life industrial goods: cable leftovers, electronic scrap, scrap from automobiles and fittings with a relatively high and calculable copper content
- *Intermediate or residual materials*: such as slag, dross, electroplating sludge, etc.

Scrap can also be classified according to *composition*:
- *Genuine, clean scrap*: such as wires or cable scrap that is directly re-melted
- *Scrap* with a *certain composition ratio*: alloy scrap of brass, bronze, red brass, etc. This scrap can be treated by certain pyrometallurgical processes (smelting or fire refinement).
- *Scrap*: of which *the metallic components to be separated* are *combined with other metallic and/or non-metallic components*. This includes electric motors, automobile scrap, electronic scrap, etc. The automotive industry essentially deals with this scrap, which is subject to scrap processing by law.

The first stage of scrap preparation is mechanical crushing and sorting. The most important processes or aggregates for this process are shredding, magnet separator, air separator and floating–sinking installations. Extraction of non-ferrous metals initially involves dismantling cars by hand, which is economical for some car parts such as starter motors, alternators, cable harnesses, central electronic systems, air conditioning, heat exchangers, catalysts, etc. Some parts must be dismantled because of statutory provisions.

The vehicle is then reduced in a shredder to car scrap in pieces of 100 to 250 mm. Separation of the mixture of materials into light and high-ash components is performed by an air separator and the separation of iron material from the high-ash component is carried out in a magnetic separator. There are various processes for separating and sorting various metals, eddy current separation, electrostatic sorting, identification by way of laser pulses, etc. The sorted copper-containing materials are processed by recycling smelters. High quality scrap, such as conductors and wires, must also be processed beforehand, e.g. by cryogenic separation of the embrittled synthetic insulation. Electronic scrap consisting essentially of composite material (copper-clad synthetics) and several electronic components with the most varied metals is given to companies specialising in handling electronic scrap.

2.6 Environmental Protection [12]

The selection of materials based on aspects of environmental sustainability is discussed nowadays in terms of environmental balances. The more economical use of energy and resources and minimum environmental impact (or adverse effect) are considered as assessment factors. Most environmental balances are limited to the assessment of the energy consumption and a balance of materials ("life cycle inventory analysis") since the evaluation of the environmental impact by comparing various harmful substances is very complex. The life cycle assessment for copper [12] discussed here is also a pure life cycle inventory analysis. The efficiency of energy production was considered here for the energy used in all processes.

The figures show that the energy requirement for secondary copper production is only about 35 to 40% of the primary production requirement. The major part of energy in the production of pure copper, i.e. copper cathodes, is consumed in mining. Open pit mining requirements are significantly higher than those of underground mining due to the expense of extraction from barren rock. It is mainly the copper content in the ore and the requirement for the milling process that determine the energy consumption during ore preparation. If cathodes in Central Europe are produced from copper concentrates, the requirement for their transport alone is 1.9 GJ/t (0.53 kWh/kg) of copper cathodes. The processing of concentrates into copper cathodes requires 19 GJ/t (5.3 kWh/kg) excluding transport, only 30% of which is for the actual production process; 15% is for the cleaning of waste gases and a significant part is for the provision of electrical energy. Over the years the energy requirement for copper production has been reduced considerably just by process development and optimization.

Copper production is almost free of waste. Sulfuric acid and slag are also produced and marketed as by-products of primary copper production. Sulfuric acid is mainly sold to the chemical industry while slag is often used as a building material due to its good environmental compatibility. Metal oxides occur primarily in secondary copper production. They are used, for instance, as pigments in the paint industry or as a source of raw materials due to the metals they contain. Small quantities of waste, e.g. those containing arsenic from wastewater treatment, are disposed of properly.

Anode slimes occurring as a by-product of the copper electrolysis process essentially contain precious metals, selenium and tellurium. These valuable materials are extracted in marketable form in further processing steps.

Solid waste in large quantities only occurs in the field of mining as barren rock (excavation waste), which does not arise in underground mining due to the remaining overlying rock, and as processing residue. The latter is dumped as mine waste or deposited in tailings ponds. Smelting and processing copper produces almost no residual materials.

There are extremely low levels of atmospheric residual emissions in domestic copper production due to the very strict environmental directives. Atmospheric emissions are essentially due to the provision of energy by power stations and those caused by transport. The copper industry has also succeeded in mastering the problems of the formation and removal of dioxins and furans. Values of 1 ng TE/m^3 (1 ng = 0.000000001 g) are no longer exceeded and often not even approached.

3
Classification of Copper Materials

This section deals with classification of EN standardized copper materials but only goes into more detail regarding copper materials used in the construction of automobiles. As required and regarding other copper materials, reference is made to the standards stated in the Annex (Standards and Specifications). The European materials numbering system for copper and copper alloys is standardized in EN 1412. The technical specification CEN/TS 13388 (predecessor of a central materials standard) contains an overview of compositions and products of copper and copper alloys published as a pre-standard in September 2004. This pre-standard itemizes all copper materials with their compositions and the corresponding product specifications are listed. The standards that are of interest here are compiled in the Annex.

3.1
Wrought Copper Materials

Wrought copper materials include all copper materials processed by massive forming such as extrusion, roll casting rolling, drawing or forging into semi-finished products (rods, profiles, wires, sheets, plates, strips or parts). The compositions of wrought copper materials are also stated in the respective product specifications. These specifications also contain mechanical properties, dimensions, tolerances and surface characteristics as far as requirements are concerned. They also contain details of definitions, designations and ordering information and information on test pieces, test procedures, declarations of conformity and test certificates besides markings, packaging and labelling.

3.1.1
Copper [1, 14]

Besides pure copper, oxygen-containing copper grades, containing controlled quantities of oxygen in the form of copper(I)oxide, and oxygen-free copper types with residual contents of deoxidising agent (preferably phosphorus) are normally described as "copper". Wet or pyrometallurgically refined wrought materials have a purity of at least 99.90% Cu and oxygen-free, non-deoxidised types

Copper in the Automotive Industry. Hansjörg Lipowsky and Emin Arpaci

ISBN: 978-3-527-31769-1

Table 2 A selection of wrought copper types (unalloyed or silver-bearing) for automotive engineering.

According to EN			Designation according to		Notes on properties and applications
Code	Material No.	Composition[a]	ASTM[a] (UNS)	JIS[a]	
Unalloyed copper types					
Cu-ETP	CW004A	Cu[b] 99.90 (min) O[c] 0.040 (max)	C11000	C1100	Conductivity ≥58 mΩ/mm², very difficult to weld and braze
Cu-FRHC	CW005A	Cu[b] 99.90 (min) O[c] 0.040 (max)	C11020	–	
Cu-OF	CW008A	Cu[b, d] 99.95 (min)	C10200	C1020	Conductivity ≥58 mΩ/mm², weldable and brazeable
Phosphorus copper types					
Cu-PHC	CW020A	Cu[b, d] 99.95 (min) P 0.001–0.006	C10300	–	High conductivity, easily weldable and brazeable
Cu-HCP	CW021A	Cu[b, d] 99.95 (min) P 0.002–0.007		–	
Cu-DHP	CW024A	Cu[b, e] 99.90 (min) P 0.015–0.040	C12200	C1220 C1221	No conductivity requirement; easily weldable and brazeable
Silver-bearing copper types					
CuAg 0.10	CW013A	Ag 0.08–0.12 Cu Rest	C11600	–	Contact members, collector rings
CuAg 0.10P	CW016A	Ag 0.08–0.12 P 0.001–0.007 Cu[d] Rest	C10700	–	As above, easily solderable and weldable

a) The tolerance ranges of the composition of the copper types standardized in other countries are not always consistent with findings according to EN.
b) Including silver up to max 0.015%.
c) Oxygen content up to 0.060% is permissible if agreed between buyer and supplier.
d) The manufacturer must set the oxygen content in such a way that the material satisfies the demands on hydrogen resistance according to EN 1976.
e) If necessary, the total of the other elements, apart from silver and phosphorus must be agreed between buyer and supplier.

and some phosphorus copper types are as high as 99.95% pure (if not higher). The chemical composition is in CEN/TS 13388 or the respective product specification. Table 2 shows the qualities of copper used for automotive engineering.

The unalloyed copper types Cu-ETP and Cu-FRHC are used exclusively in wire form when under high demands of electrical conductivity. Both types are very difficult to weld or braze. The oxygen-free types are used for welding and brazing, so Cu-OF is used for electronic applications and is free of evaporable elements.

The phosphorus copper types Cu-PHC and Cu-HCP are used when greater electrical conductivity with increased demands of malleability and weldability and brazeability are placed on semi-finished products. Cu-DHP is the material for general applications with no demands on conductivity.

The silver-bearing copper material CuAg 0.10 is highly conductive, in comparison with highly conductive copper types, has an improved tempering resistance and better creep resistance at higher temperatures and is used for example for collector rings and contacts. The easily brazeable and weldable CuAg 0.10 P has the same properties and applications as CuAg 0.10.

3.1.2
Low Alloyed Wrought Copper [14, 15]

This alloy group contains copper alloys in which the properties of pure copper, e.g., strength, softening temperature and machinability are improved by small additions of various alloying elements, up to a maximum of 5% (CEN/TS 13388) for which one has to accept a certain deterioration of some properties, for example conductivity. A general distinction (not according to the specification) must be made between non-age-hardenable and age-hardenable alloys. This means a distinction must be made between alloys that can only be hardened by cold working and those that can also be hardened by heat treatment.

In the case of non-age-hardenable alloys, additions of e.g. silver, iron and magnesium serve to increase strength and particularly the softening temperature and therefore the tempering resistance. Silver, iron and magnesium are particularly good alloying elements in cases where high strength and conductivity are required. Machinability is increased by adding sulfur, lead or tellurium as a chip breaker.

By adding beryllium, nickel and silicon, zirconium or chromium and zirconium, age-hardenable alloys are produced, which, following heat treatment, possess not only increased strength but also higher conductivity. Table 3 contains a selection of standardized materials and their composition used in automotive engineering (according to CEN/TS 13388).

In automotive engineering, low alloyed copper materials are mainly used for electrotechnical purposes. Examples are commutator plates, contact members and lead frames. The conductivity, which is still good, is combined with high strength. Springs of age-hardenable alloys in automotive safety devices have also been proven, as for example springs of copper beryllium have a long service life and are completely maintenance-free.

Table 3 A selection of low alloyed wrought copper for automotive engineering.

According to EN			Designation according to		Notes on properties and applications
Code	Material No.	Composition[a)]	ASTM[a)] (UNS)	JIS[a)]	
Non-age-hardenable low alloyed wrought copper					
CuFe2P	CW107C	Fe 2.1–2.6 Zn 0.05–0.20 P 0.015–0.15 Cu Rest	C19400	–	lead frames
CuSP	CW114C	S 0.2–0.7 P 0.003–0.012 Cu Rest	C14700	–	
CuTeP	CW118C	Te 0.4–0.7 P 0.003–0.012 Cu Rest	C14500 C14510 C14520	–	free machining copper
CuZn0.5	CW119C	Zn 0.1–1.0 Cu Rest	–	–	lead frames
Age-hardenable low alloyed wrought copper alloys					
CuBe2	CW101C	Be 1.8–2.1 Cu Rest	C17200	C1720	high-strength springs
CuBe2Pb	CW102C	Be 1.8–2.0 Pb 0.2–0.6 Cu Rest	C17300	–	as above but better machinable
CuCo2Be	CW104C	Co 2.0–2.8 Be 0.4–0.7 Cu Rest	C17500	–	springs
CuNi2Si	CW111C	Ni 1.6–2.5 Si 0.4–0.8 Cu Rest	C64700	–	high tensile strength bolts
CuCr1Zr	CW106C	Cr 0.5–1.2 Zr 0.03–0.3 Cu Rest	C18150	–	contact members
CuZr	CW120C	Zr 0.1–0.2 Cu Rest	C15000	–	wires, springs, contact members

a) The tolerance ranges of the composition of the alloys standardized in other countries are not always consistent with findings according to EN.

3.1.3
Wrought Copper–Zinc Alloys [14, 16]

The most common wrought copper–zinc alloys contain 5 to 45% zinc besides copper. The term "brass" is generally used for alloys of the base metal copper with zinc. Copper–zinc–lead alloys contain up to 3.5% lead to improve the machining properties. Multi-copper–zinc alloys with further additions alone or in combination with e.g. aluminum, iron, manganese, nickel, silicon and tin have up to now been called "special brass". The above-mentioned classification was maintained in binary, leaded and in copper–zinc alloys with additives in CEN/TS 13388. Table 4 shows the most important copper–zinc alloys used in automotive engineering.

Binary alloys can be particularly easily cold formed and are used for example for connectors in automotive engineering. The binary material CuZn37 can be easily welded (keep zinc-evaporation in mind) and brazed, is the least expensive and is therefore the main alloy for cold formed parts.

Of the group of leaded copper–zinc alloys, CuZn36Pb3 can be very easily machined and be cold formed to a certain extent. The former free-cutting brass with 57 to 60% Cu and 1.5 to 3.5% Pb was divided into three alloy grades, in a drilling and lathing quality (CuZn39Pb2), into the best free-machining alloy with 3% Pb (CuZn39Pb3) and an alloy grade preferred for hot stamping (CuZn40Pb2).

The multi-alloy CuZn31Si1 is very wear resistant and has been shown to be useful in producing wrapped bushings for the automotive industry. CuZn38-Mn1Al shows similar properties for medium stress with good corrosion resistance. CuZn37Mn3Al2PbSi has the highest levels of strength with greater amounts of Al and Mn and additions of Si; this material can be easily hot formed but is not easily cold formed. It is used in automotive engineering for oscillating bearings and synchronizer rings.

3.1.4
Wrought Copper–Tin Alloys [14, 17]

The term "copper–tin alloys" includes alloys of copper with the main alloying element tin. These alloys are also correctly referred to as "bronze" or "tin bronze". The term "bronze" should not be used for other copper alloys. Wrought copper–tin alloys contain 1.5 to 9.0% tin according to CEN/TS 13388. Phosphorus is added to these alloys in most cases for deoxidation (ensured by residual content of $>0.01\%$ P). Phosphorus is however also consciously added as an alloying element; such alloys are described as a distinction from tin bronze as "phosphor bronze". For economic reasons, some alloys also contain zinc as well as tin. Table 5 contains a selection of wrought copper-tin alloys for automotive engineering and their composition and applications.

Wrought copper–tin alloys are easily work-hardened by cold forming, are corrosion resistant and sufficiently conductive, they are preferred for spring strips

Table 4 A selection of wrought copper–zinc alloys for automotive engineering.

According to EN			Designation according to		Notes on properties and applications
Code	Material No.	Composition[a]	ASTM[a] (UNS)	JIS[a]	
Binary copper–zinc alloys					
CuZn15	CW502L	Cu 84.0–86.0 Zn Rest	C23000	C2300	can be very easily cold formed, spring strips
CuZn30	CW505L	Cu 69.0–71.0 Zn Rest	C26000	C2600	can be very easily cold formed, spring parts for connectors, radiator strips, leaf springs
CuZn33	CW506L	Cu 66.0–68.0 Zn Rest	C26800	C2680	
CuZn37	CW508L	Cu 62.0–64.0 Zn Rest	C27400	C2720	main alloy for cold forming, easily solderable and weldable
Copper–zinc–lead alloys					
CuZn36Pb3	CW603N	Cu 60.0–62.0 Pb 2.5–3.5 Zn Rest	C35600 C36000	C3601 C3602	free-machinable and cold formable
CuZn39Pb2	CW612N	Cu 59.0–60.0 Pb 1.6–2.5 Zn Rest	C37700	C3771 C3713 C3561	can be easily free-machined, drilled, milled and punched
CuZn39Pb3	CW614N	Cu 57.0–59.0 Pb 2.5–3.5 Zn Rest	C38500	C3561 C3603 C3604	free-machinable, main free-machining alloy, profile turned parts
CuZn40Pb2	CW617N	Cu 57.0–59.0 Pb 1.6–2.5 Zn Rest	C38000	C3561 C3603 C3604 C3771	free-machinable, easily hot formable, extruded profiles

Table 4 (continued)

According to EN			Designation according to		Notes on properties and applications
Code	Material No.	Composition[a)]	ASTM[a)] (UNS)	JIS[a)]	
Copper–zinc alloys, binary alloys					
CuZn31Si1	CW708R	Cu 66.0–70.0 Si 0.7–1.3 Zn Rest	C87900	–	for sliding purposes, bushings, guides
CuZn38Mn1Al	CW716R	Cu 59.0–61.5 Mn 0.6–1.8 Al 0.3–1.3 Zn Rest	C86500	C6782	construction material of medium strength for sliding purposes
CuZn37Mn3Al2PbSi	CW713R	Cu 57.0–59.0 Mn 1.5–3.0 Al 1.3–2.3 Pb 0.2–0.8 Si 0.3–1.3 Zn Rest	C67400	–	construction material of high strength for sliding purposes, synchronizer rings

a) The tolerance ranges of the composition of the alloys standardized in other countries are not always consistent with findings according to EN.

Table 5 A selection of wrought copper-tin materials for automotive engineering.

According to EN			Designation according to		Notes on properties and applications
Code	Material No.	Composition[a)]	ASTM[a)] (UNS)	JIS[a)]	
CuSn4	CW450K	Sn 3.5–4.5 P 0.01–0.4 Cu Rest	C51100	C5111	connectors, conductive springs
CuSn6	CW452K	Sn 5.5–7.0 P 0.01–0.4 Cu Rest	C51900	C5191	springs, connectors, wires, tubes and spring tubes
CuSn8	CW453K	Sn 7.5–8.5 P 0.01–0.4 Cu Rest	C52100	C5210 C5212	sliding elements, thin-walled bushings and strips

a) The tolerance ranges of the composition of the alloys standardized in other countries are not always consistent with findings according to EN.

or wires, e.g. for relay springs and high quality connectors and also for conductive springs. These materials also show good sliding properties. The alloy CuSn8 has shown itself to be wear resistant for mechanically highly stressed, thin-walled bushings at high speed.

3.1.5 Wrought Copper–Nickel Alloys [14, 18]

Copper materials with copper as the base metal and nickel as main alloying element with or without other alloying elements are described as copper–nickel alloys. Copper and nickel form an uninterrupted series of solid solutions so that alloys are theoretically possible in all Cu-/Ni-mixing ratios. The wrought copper–nickel alloys show a Ni-content of 8.5 to 32.0% according to CEN/TS 13388. Table 6 shows a selection of alloys and their composition important for automotive engineering.

The material CuNi9Sn is easily cold formable and air corrosion resistant. When the spring is hard, it possesses very good stress relaxation properties and is used for resilient contacts in relays, switches and connectors. The material CuNi10Fe1Mn is highly resistant to corrosion, cavitation and erosion and is used for brake pipes and intercoolers (charge air coolers). As resistor alloy CuNi44Ni1 shows a very high constancy of electric resistance under varying temperatures.

3.1.6 Wrought Copper–Nickel–Zinc Alloys [14, 19]

This term describes copper alloys containing nickel and zinc as main alloying elements. These alloys are described as "nickel silver" due to their silver-like color. They contain between 42 and 66% Cu and from 6 to 19% Ni (Rest Zn) according to CEN/TS 13388. Like brass, up to 3.3% lead as chip breaker is added to enhance turning and drilling qualities. Only a few lead-free alloys with higher strength and better corrosion resistance than brass are of technical importance for the automotive industry; they are used in strip form above all for contact springs in electrical relays. Table 7 shows a selection of materials relevant here together with their composition.

3.1.7 Wrought Copper–Aluminum Alloys [14, 20]

Copper–aluminum alloys are alloys of copper with the main alloying element aluminum which are sometimes also known as "aluminum bronze". They have Al contents of 4.0 to 12.5% according to CEN/TS 13388. These alloys can also contain e.g. iron, nickel, manganese or silicon to improve other properties. Provided there is very good lubrication, wrought copper–aluminum alloys have proved efficient as sliding materials, based on their high strength and wear re-

Table 6 A selection of wrought copper–nickel materials for automotive engineering.

According to EN			Designation according to		Notes on properties and applications
Code	Material No.	Composition[a)]	ASTM[a)] (UNS)	JIS[a)]	
CuNi9Sn2	CW351H	Ni 8.5–10.5 Sn 1.8–2.8 Cu Rest	C72500	–	spring contacts in relays, switches and connectors
CuNi10Fe1Mn	CW352H	Ni 9.0–11.0[b)] Fe 1.0–2.0 Mn 0.5–1.0 Cu Rest	C70600	C7060	brake pipes

a) The tolerance ranges of the composition of the alloys standardized in other countries are not always consistent with findings according to EN.

b) Co max 0.1% is counted as Ni.

Table 7 A selection of wrought copper–nickel–zinc materials (nickel silver) for automotive engineering.

According to EN			Designation according to		Notes on properties and applications
Code	Material No.	Composition[a)]	ASTM[a)] (UNS)	JIS[a)]	
CuNi18Zn20	CW409J	Cu 60.0–63.0 Ni 17.0–19.0 Zn Rest	C75200	C7941	easily cold formable, spring material, spring strips
CuNi18Zn27	CW410J	Cu 53.0–56.0 Ni 17.0–19.0 Zn Rest	C77000	C7701	

a) The tolerance ranges of the composition of the alloys standardized in other countries are not always consistent with findings according to EN.

sistance, for highly and shock-loaded sliding elements, for bushings and worm gears. Until now, these materials have hardly been used in automotive engineering.

3.2 Copper Casting Materials

Copper casting materials include all those materials suitable for producing parts by casting. The prepared casts solidify during the casting process close to the final dimension and are subsequently just cleaned and only in some cases, e.g. for functional surfaces, machined. They can be produced by sand, chill, centrifugal, continuous or die casting. The suitability of the materials for the different casting processes is contained in EN 1982 by identification numbers attached to the material designation. These are listed in Table 8.

As for wrought copper materials, the compositions of the copper casting materials are contained in CEN/TS 13388 "Technical specifications" (published as a pre-standard in September 2004) with a classification according to alloy groups and the respective suitable casting procedures. Besides the chemical composition, EN 1982 also shows the requirements for mechanical properties, electrical conductivity, microstructure and grain size as well as dezincification resistance and freedom of defects. They also contain details on definitions, designations and ordering information, information on test pieces, test procedures, declarations of conformity and test certificates as well as markings, labelling and packaging.

3.2.1 Copper and Copper–Chromium Casting Materials [1, 21]

The standardization includes copper casting and copper–chromium casting alloys. Sand casting and permanent mold casting are standardized for both materials. They have been listed in Table 9 (details according to CEN/TS 13388 and EN 1982).

The unalloyed copper cast Cu–C is used in automotive engineering for conduction purposes, in particular, power-conducting parts. A comparison has been made between the types of wrought copper and those of a lower copper content not laid down in the standardization. The attainable values for electrical conduc-

Table 8 Identification of casting procedures for copper casting alloys in EN 1982.

Casting procedure	Sand casting	Permanent mold casting	Centrifugal casting	Continuous casting	Pressure die casting
Identification	GS	GM	GZ	GC	GP

Table 9 Copper casting materials; copper and copper–chromium alloys.

According to EN					Designation according to		Notes on properties and applications
Code	Material No.	Composition[a]	Casting procedure and designation[b]		ASTM[a] (UNS)	JIS[a]	
Cu-C	CC040A[c]	not determined	permanent mold casting-GM		C80100 C80410	CuC2 CuC3	power conducting parts, conductivity min. 55 MS/m
			sand casting-GS	Grade A Grade B Grade C[d]	C81100	CuC1	power conducting parts, minimum conductivity for grade A 50, grade B 45 and grade C 32 MS/m
CuCr1-C	CC140C[c]	Cr 0.4–1.2 Cu Rest	permanent mold casting-GM sand casting-GS		C81500	–	application as above, higher strength than Cu-C, conductivity min. 45 MS/m[e]

a) The tolerance ranges of the composition of the materials standardized in other countries are not always consistent with findings according to EN.
b) Classification according to casting procedure refers only to EN materials and not materials according to ASTM or JIS.
c) Ingots of this material are not laid down in EN 1982.
d) This type refers to certain heat transfer applications e.g. water-cooled hoisting appliances.
e) When completely annealed.

tivity are also lower. Permanent mold casting at 55 MS/m has a higher electrical conductivity than sand casting, the electrical conductivity of which is between 32 and 50 MS/m. The tensile strength of copper castings is min. 150 N/mm^2 (with 0.2% proof stress 40 N/mm^2 and an elongation of min. 25%).

The copper–chromium casting alloy is included in the above specifications as an age-hardenable low alloy copper casting material. This material has a high tensile strength ≥300 N/mm^2, a 0.2% proof stress of 200 N/mm^2 and an elongation of min. 10% with a minimum of 45 MS/m (when fully annealed). It is also used in automotive engineering for conductive parts.

For copper or copper–chromium castings, it is to be stated when ordering, which grade (for sand casting, A, B or C) is required, whether electrical conductivity is to be determined and if test details and test specimens are to be provided.

3.2.2 Copper–Zinc Casting Alloys [16, 21]

For copper–zinc casting alloys (also called cast brass), a general distinction can be made between leaded materials and other casting materials containing other alloying elements such as iron, aluminum, manganese and silicon. Copper–zinc casting alloys are also standardized in CEN/TS 13388 and EN 1982 just like other copper casting materials.

These materials are only used to a small extent in automotive engineering. Table 10 lists some typical copper–zinc casting alloys with their compositions. Battery terminals and shift forks of leaded brass produced by die casting or permanent mold casting are examples of their application in automotive engineering. Smaller, often also machined, gravity die cast parts are used for various components (machines and electrical engineering).

3.2.3 Copper–Tin Casting Alloys [21, 22]

Copper–tin casting alloys (standardized in CEN/TS 13388 and EN 1982) contain 9 to 13% tin and are also described as cast tin bronze. They often also contain up to 2% nickel to reduce the influence of the wall thickness on the mechanical properties. The lead contents improve the sliding and machining properties and increase the density of the cast structure.

These cast materials are used for plain bearings in e.g. automobile steering gears and worm gears. The lead-free nickel-bearing alloy CuSn12Ni2-C-GZ or -GC is used in the latter case. The centrifugal and continuous casting qualities listed in the specification are, in principle, preferred processes as they show improved mechanical and sliding properties. Table 11 shows a selection of materials of interest to the automotive industry.

Table 10 A selection of copper–zinc casting alloys for automotive engineering.

According to EN			Designation according to		Notes on properties and applications
Code	Material No.	Composition[a]	ASTM[a] (UNS)	JIS[a]	
CuZn38Al-C	CC767S	Cu[b] 59.0–64.0 Al 0.1–0.8 Zn Rest	C85700	YbsC3 YbsCIn3	complex engineering parts, mainly for electronic technology
CuZn35Pb2Al-C[c]	CC752S	Cu 61.5–64.5 Pb 1.5–2,5 Al 0.3–0.70 Zn Rest[d, e]			machinable parts, shift forks
CuZn16Si4-C	CC761S	Cu 78.0–83.0 Si 3.0–5.0 Zn Rest	C87500	SzBC2 SzBC3	high-strength parts for electronic technology

a) The tolerance ranges of the composition of the alloys standardized in other countries are not always consistent with findings according to EN.
b) Including nickel.
c) Castings of this alloy must satisfy requirements on dezincification resistance according to 6.5 of EN 1982.
d) Sb and As are alternative dezincification inhibitors. If Sb is added as an inhibitor, As content must be limited to 0.04%. If As is added as an inhibitor, Sb content may not exceed max. 0.04%.
e) (Sb + As) may not exceed 0.15%.

Table 11 A selection of copper–tin casting alloys for automotive engineering.

According to EN			Designation according to		Notes on properties and applications
Code	Material No.	Composition[a)]	ASTM[a)] (UNS)	JIS[a)]	
CuSn12Ni2-C	CC484K	Cu 84.5–87.5 Sn 11.0–13.0 Ni 1.5–2.5	C91700	PBCIn2	worm gears
CuSn11Pb2-C	CC482K	Cu 83.5–87.0 Sn 10.5–12.5 Pb 0.7–2.5	C92500	LBC2	high capacity plain bearings

a) The tolerance ranges of the composition of the alloys standardized in other countries are not always consistent with findings according to EN.

Table 12 A selection of copper–tin–lead casting alloys for automotive engineering.

According to EN			Designation according to		Notes on properties and applications
Code	Material No.	Composition[a)]	ASTM[a)] (UNS)	JIS[a)]	
CuSn10Pb10-C	CC495K	Cu[b)] 78.0–82.0 Sn 9.0–11.0 Pb 8.0–11.0	C93700	LBC3 LBCIn3 LBC3C	plain bearing, composite bearing
CuSn7Pb15-C	CC496K	Cu[b)] 74.0–80.0 Sn 6.0–8.0 Pb 13.0–17.0 Ni 0.5–2.0	C93800 C93900	LBC4 LBCIn4 LBC4C	plain bearing, composite bearing
CuSn5Pb20-C	CC497K	Cu[b)] 70.0–78.0 Sn 4.0–6.0 Pb 18.0–23.0 Ni 0.5–2.5	C94100	–	highly loaded composite bearing
CuSn7Zn4Pb7-C	CC493K	Cu[b)] 81.0–85.0 Sn 6.0–8.0 Zn 2.0–5.0 Pb 5.0–8.0	C93200	–	moderate capacity plain bearing

a) The tolerance ranges of the composition of the alloys standardized in other countries are not always consistent with findings according to EN.
b) Including nickel.

3.2.4 **Copper–Tin–Lead Casting Alloys** [21–23]

Copper–tin–lead casting alloys are standardized in CEN/TS 13388 and EN 1982. These alloys can contain from 2 to 11% tin and 1 to 23% lead. Copper–tin–lead casting alloys without added zinc are also called casting tin–lead bronze. Materials of this group that may contain between 1.5 and 10% zinc as an alloying element besides tin and lead are also known under the term red brass. Copper–tin–lead casting alloys normally also have a small amount of nickel added.

Copper–tin–lead casting alloys are of technical importance in automotive engineering only as sliding materials. Continuous or centrifugal casting is nearly always used for producing piston pin and rocker arm bushings with 10 or 15% lead. In contrast, copper–tin–lead casting alloys with lead contents of 18 to 23% are nearly always produced by the strip steel process. Wrapped bushings for crankshafts with an additional layer of high-lead alloy (overlay) are used as so-called three-component bearings in combustion engines. Table 12 shows a selection of the most important casting alloys with their compositions.

3.2.5 **Copper–Aluminum Casting Alloys** [20, 21]

The customary copper–aluminum casting alloys standardized in CEN/TS 13388 and EN 1982 have a heterogeneous microstructure and contain 8 to 12% aluminum and other alloying elements such as iron and nickel besides copper as base element. They are also known as casting aluminum bronze.

Parts produced for the transport industry by gravity die casting and parts for plain bearings and sliding elements produced by continuous and centrifugal casting are of technical importance. Table 13 contains some examples of the important copper–aluminum casting alloys with their compositions. The iron-containing copper–aluminum casting alloy, due to its high toughness, is used for plain bearings with very high impact loads, bevel gears, synchronizer rings, switching segments and shift forks. The standardization lists three copper–aluminum casting alloys with iron and nickel with different strength levels (Table 13 lists just one alloy variant). These bearings are also used for fully lubricated plain bearings with high impact loads, e.g. crankshafts and toggle bearings, worm gears and helical gears.

3.2.6 **Copper–Manganese–Aluminum and Copper–Nickel Casting Alloys** [18, 21]

Copper–manganese–aluminum casting alloys and copper–nickel casting alloys (nickel content from 9 to 31%) are standardized in CEN/TS 13388 and EN 1982. These materials have nevertheless failed until now to gain any technical importance for automotive engineering.

Table 13 A selection of copper–aluminum casting alloys for automotive engineering.

According to EN			Designation according to		Notes on properties and applications
Code	**Material No.**	**Composition**[a)]	**ASTM**[a)] **(UNS)**	**JIS**[a)]	
CuAl10Fe2-C[b)]	CC331G	Cu 83.0–89.5 Al 8.5–10.5 Fe 1.5–3.5	C95200	AlBC1 AlBCIn1	bevel gears, synchronizer rings, switching segments and shift forks
CuAl10Fe5Ni5-C	CC333G	Cu 76.0–83.0 Al 8.5–10.5 Fe 4.0–5.5[c)] Ni 4.0–6.0[c)]	C95800	AlBC3 AlBCIn3	highly stressed crankshaft- and toggle bearings, worm gears and helical gears

a) The tolerance ranges of the composition of the alloys standardized in other countries are not always consistent with findings according to EN.

b) For castings to be welded, lead content must be limited to 0.03%.

c) For permanent mold casting, the minimum iron content for ingots and castings is 3.0% and the minimum nickel content 3.7%.

3.3 Composites

Until now, copper composites have been used sporadically in automotive technology. The various composites should be briefly treated taking account of future development potential. A difference is to be made between composites manufactured by various production processes:

- Dispersion-hardened copper
- Composites consisting of at least two different metals and produced by plating (or weld cladding), recasting or sintering.

Dispersion-hardened copper consists of a copper matrix with an incorporated component of one or more other materials. Even though there have been various attempts to harden copper by fibers or dispersion of solid components without affecting the conductivity too much, practically only one copper, dispersion-hardened by aluminum oxide particles, under the trade name "**Glidcop**" has been commercially available until now. This dispersion-hardened copper is produced with 0.3, 0.5 and 1.1 wt.% aluminum oxide by a patented internal oxidation process. By this process the finest oxide particles distributed uniformly in the material can be produced. According to the aluminum oxide content, the material, compared with highly conductive copper, has an electrical conductivity of 78 to 92% and, depending on the semi-finished product, it has twice or more than twice the strength. As the hardener in the copper matrix is thermally stable aluminum oxide, the material can be used quite close to the melting point. In this way, it will retain higher strength linked with greater conductivity, in comparison with copper or low alloy copper types, which soften when reaching a limiting temperature (re-crystallization temperature). Regarding strength, combined with higher electrical and thermal conductivity and insensitivity to softening at high temperatures, Glidcop is superior to any other copper material.

Plated materials, e.g. steel or aluminum with one- or two-side copper or copper alloy plating are commercially available. The platings consist of roll- or explosive plating. Examples are "CUPAL" power rails of copper-clad aluminum (to facilitate the electrical connection between aluminum and copper conductors). Weld cladding other materials with copper materials is often carried out for reasons of wear or corrosion protection.

Parts of other metals, e.g. cooling pipes, can be cast in copper or copper alloy castings. The strip steel process, in which copper alloys with for example 10% lead and 10% tin or 22% lead and 15% tin are continuously cast onto steel strips, is of technical importance in automotive engineering. Bearings and bushings with steel back are wrapped with the bimetal and other parts are produced for combustion engines. For example, a copper–tin alloy as sliding material is also poured onto control discs for axial piston pumps.

There are also composites where a porous layer of copper or copper alloy is initially sintered onto a steel strip to subsequently infiltrate this layer with lead, a lead alloy, graphite or polymers. Bushings consisting of these composites have the strength of the steel and the specific sliding properties of the applied material.

3.4 Powder Metallurgical Materials [24]

Powder metallurgically produced (PM) materials can be classified into conventional sintered materials and powder metallurgically produced dispersion-hardened copper types. Strictly speaking, "Glidcop" belongs in this group. Until now, there are no EN specifications for materials in this group or for conventionally sintered materials.

PM materials are most important in the application of "self-lubricating plain bearings" of sintered bronze. This is based on a powder mixture of 90% copper powder (electro-deposited or atomised) and 10% tin powder. The powder is only compressed so as to leave a pore volume of approximately 25%. After sintering and calibration the bushings are infiltrated with lubricating oil. Copper or copper alloy powder is also used for friction linings.

The high-leaded copper–tin-lead-alloys tend to suffer lead-segregation and are difficult to cast, therefore steel–lead-bronze strips have been developed by sintering the lead bronze onto the steel strip. This strip is also used to produce wrapped three-component plain bearings. This procedure is however losing popularity compared with the strip steel process as the latter has been further developed in the meantime and plain bearings with cast lead–tin alloys are more stress resistant than bearings with sintered lead bronze.

4 Wrought Copper Materials [14]

The largest share of wrought copper materials is shipped to the manufacturing industry in the form of semi-finished products.

4.1 Manufacture of Semi-finished Products

The production process of semi-finished products can be roughly classified into the steps smelting and casting, hot and cold forming, often connected with heat treatment and final processing.

4.1.1 Smelting and Casting

The manufacture of semi-finished products begins with smelting the copper materials and casting these into certain formats e.g. slabs, billets etc. Smelting and casting takes place according to a production program set out precisely in advance. Both the use of metal and adherence to the alloy composition are strictly monitored. A specimen of the melt is sent by pneumatic delivery to an analysis machine (spectrum analysis) so that within two minutes the result is available in the smelting shop and the chemical composition can be corrected just before casting. Smelting normally takes place in electric induction furnaces and the formats are cast in modern continuous casting plants. Slabs are preferably produced by semi-continuous casting, billets however by continuous casting.

4.1.2 Hot Forming

Hot forming of copper materials (like any other metal) is carried out above the re-crystallization temperature. Recrystallization temperatures for copper and copper alloys are between 650 and 950 °C depending on the material composition. No work-hardening can take place on hot forming because of recrystallization; rather, work-hardening is permanently reduced by immediate (*in situ*) re-

Copper in the Automotive Industry. Hansjörg Lipowsky and Emin Arpaci

ISBN: 978-3-527-31769-1

crystallization (soft annealing). Furthermore, resistance to deformation of the metal is generally reduced with increasing temperature.

The various copper materials have different hot forming properties. With the homogeneous single-phase (α-phase) copper alloys, resistance to deformation increases, starting from pure copper, with increasing content of alloying elements. The heterogeneous alloys of the Cu–Zn or Cu–Al system, however, have a lower resistance to hot forming; the newly appearing β-mixed crystal is more easily hot formable than the α-mixed crystal. Brass is therefore particularly characterized in the β-phase range by a very low resistance to hot deformation. The best hot forming properties in the sequence of suitability are attributed to heterogeneous brass with at least 37% Zn content, unalloyed copper and also CuAl alloys, while CuSn, CuSnPb, CuNiZn and CuNi alloys are not so easily hot formable.

4.1.3
Cold Forming

Yield strength rises with increasing deformation (work-hardening) in the case of cold forming, which takes place below the recrystallization temperature. The homogeneous copper alloys are the best cold formable after unalloyed copper because the α-mixed crystal shows the best cold forming properties. A rule of thumb is that, with the exception of copper–nickel alloys, the work hardening capacity of copper materials is determined by the relative hardness in the soft state. The material's strength increases and its elongation, and therefore also its further cold-formability, decreases with increasing cold working.

Pure unalloyed copper and the copper-rich copper–zinc alloys are excellently cold formable but copper–nickel alloys, most low alloyed copper alloys and copper–tin alloys are also relatively easily cold formable. Copper–aluminum, highly leaded or highly tin-bearing copper–tin–lead alloys and also copper–zinc-alloys with high tin content, the β- and (α+β) brasses, are less easily cold formable.

4.1.4
Production of Sheet and Strip

Stock material for rolling sheets and strips is cast plates from semi-continuous casting plants. These are approximately 5 m long, 600 to 800 mm wide and 120 mm thick. In the case of materials that are difficult to hot roll, e.g. copper–tin alloys, as well as for smaller lot sizes for commercial reasons for other copper materials, production often starts from a horizontally continuously cast strip, e.g. 600 mm wide and 25 mm thick with direct cold rolling. In this case, the otherwise necessary steps of "heating" and "hot rolling" are obsolete.

For hot rolling, the slabs are heated to above recrystallization temperature. The slabs are then rolled down to about one tenth of their original thickness. A few tenths of a millimeter are milled off each side to remove the cast surface and scale.

The cogging strips, approximately 10 mm thick, are cold rolled in several passes to thicknesses of 2 to 3 mm according to their cold formability. One or several intermediate annealings combined with pickling and drying are required for greater thicknesses of copper materials which are difficult to roll. Thinner strips are intermediately annealed (recrystallization) under protective gas in a continuous floating furnace so pickling can be eliminated. The work hardening will be undone and cold formability regained with intermediate annealing. This is followed by final rolling passes on reversing multi-roll mills for narrow strips and copper materials difficult to roll.

4.1.5 Wire Production

Above all, wires and cords of unalloyed copper (wire harness) are of technical importance in automotive engineering. We shall therefore only go into the production of copper wires in this section. The highest portion of copper in a car is wire for electric purposes.

Roll-cast wire has established itself as the stock material of today. A process is used in which copper cathodes are melted and cast into wires according to the SOUTHWIRE or HAZELETT procedure and rolled down into wires in one heat to e.g. 12 mm ∅. The coil weights are thus normally between 5 and 8 tonnes. This saves energy previously required for re-heating wire rods.

The roll-cast wire is normally drawn, without intermediate annealing, on rough and fine drawing machines to the final dimensions. The drawing machines are multi-draw with several draws, one behind the other. Drawing speeds of the final passes are very high as large cross section reductions take place. Coiling speeds are 40 to 60 m/s. The wire is often delivered in a hard condition as it must be annealed anyway for producing lacquered wire. If annealing is necessary, it is carried out under protective gas so that pickling can be eliminated.

4.1.6 Production of Rods and Profiles

The production of rods and profiles is based on continuous cast billets with diameters of approximately 150 to 300 mm ∅ and lengths of approximately 200 to 800 mm. These are heated in induction or gas furnaces to an extrusion temperature above the recrystallization threshold. The heated bolts are then pressed in an extrusion press into a suitable profile. The shape of the strand is determined by the opening of the tool on the discharge side of the press. Large strand sections and profiles are left in straight, stretched lengths; smaller profiles are coiled.

The surfaces of the rods or extruded profiles oxidize due to the high temperature during pressing. The oxide layer of straight profiles is removed by immersion in a pickling bath; that of coiled profiles by drawing them continuously through a pickling bath. Extruded material is then generally only cold formed into the final product.

Rods from coils are continuously drawn on combination machines. In this way, the processes of drawing, testing, trimming and straightening follow one another in a single installation. Rods in straight lengths and profiles are drawn on long draw benches. In many cases however, rods with large cross sections and profiles are also delivered in the as-pressed condition after testing, trimming and straightening. Drawn rods and profiles are straightened on straighteners, the latter are also adjusted to their required profile shape. Rods and profiles are tested for freedom from defects and compliance with the shape tolerances.

4.1.7 Production of Tubes

Copper tubes can be produced by various processes. In the two most common processes, continuously cast ingots are either heated and pressed into pretubes in an extruding press or continuously cast billets are rolled in a pilger rolling mill. This is followed by a series of draws off the coil on bull blocks. Following the last draw, the coiled tubes are straightened in combined machines, eddy current tested for freedom from defects and trimmed. Soft tubes are trimmed into coils and hard tubes into straight lengths.

Tubes of copper alloy, e.g. copper–nickel alloys, are pressed on an extrusion press into relatively thin-walled pretubes using a mandrel. These tubes are then drawn on draw benches, if necessary with intermediate annealing, to final dimensions.

The production of longitudinal seam-welded tubes of copper materials has until now not established itself in Germany or the UK.

4.1.8 Forging

Unalloyed copper, $(\alpha+\beta)$ brass, wrought copper–zinc alloys with other alloying elements (special brass), copper–aluminum, a number of low alloyed wrought copper materials and the material CuNi10Fe1Mn are forgeable. These materials can be open die or closed die forged (hot stamped). Only hot stamping is of technical importance for automotive engineering.

With hot stamping, the billet (slug), normally part of a rod, is heated to forging temperature (≥ recrystallization temperature) and formed between an upper and lower die in several stages. The material is well wrought during this process. For hollow forging, the inner contours of the workpiece are pre-formed to reduce the machining cost. By pushing in steel punches from the side, the inner contours can be pre-formed to a greater extent.

Forged pieces show, compared to other procedures (e.g. casting, machining, sintering etc.) higher strength, superior surface quality and a more uniform, fine microstructure with favorable fiber orientation. Hot pressed synchronizer rings of the material CuZn37Mn3Al2PbSi have proved their value and established themselves in automotive engineering.

4.1.9
Powder Metallurgical Forming (Sintering)

For sintered compacts of copper material, suitable alloys are formed by mixing the respective powders. Prealloyed powders such as copper–zinc, copper–tin or copper–tin–lead can be used instead. Prealloyed high-leaded powder mixtures have the advantage that thorough mixing and a fine lead distribution help to prevent the exuding of lead during sintering.

The powders are mechanically or hydraulically pressed into shaped parts. Pressing at room temperature is preferred. Pressing must be carried out twice if a high degree of compression is required. The pressed parts already have a certain strength from mechanical cramping.

The parts obtain their final strength by sintering, meaning that they are heated to temperatures just below the melting point of the alloys to be sintered. Alloy components that melt below sintering temperature, such as lead, will facilitate the sintering process. The parts are calibrated, i.e. cold pressed, after sintering because pressing and sintering change the dimensions of the parts.

The main application of sintered products of copper material is for oil impregnated sintered bearings, so-called self-lubricating bearings, with a pore volume up to 30%. The pores can also be filled with other lubricants, e.g. graphite.

4.2
Shapes and Dimensions

Depending on the formability of the alloy, copper base materials are commercially available in various semi-finished product shapes and dimensions. References to available shapes and dimensions are contained in various EN specifications.

Current European specifications for semi-finished products are purely product specifications containing all copper materials for a semi-finished product for an area of applications, with distinctions being made according to different applications, e.g. general application, electrical and apparatus engineering, spring strips and wire etc. As already mentioned (see Chapter. 3), these specifications contain all details for the respective materials such as composition, mechanical properties etc. The specifications show, depending on their strength, commercially available dimensions such as thicknesses of strips, wall thicknesses of tubes, wire diameters and thicknesses, diameters or hexagonal bar sizes of rods. The specifications also contain further details on technical delivery conditions, depending on the semi-finished product type, such as limiting size, details on grain size or tolerances.

The following section shows the relevant specifications for the different semi-finished product types, sheets, strips, plates, wires, tubes, rods and extruded profiles, forgings, etc. The respective specifications are also compiled in the "Appendix – Standards and Specifications" as far as they are of importance or interest for automotive engineering.

4.2.1 Strips, Sheets and Plates [25–27]

Sheets and plates of copper or copper alloys are of hardly any technical importance for automotive engineering. On the other hand, strips of copper alloy for connectors, lead frames and conductive springs are gaining importance. Plates, sheets, strips and round blanks for general application are standardized in EN 1652. According to their definition following the standardization, plates, sheets and strips are flat rolled products, with rectangular profiles of uniform thickness. The plates are over 10 mm thick and the sheets are 0.2 to 10 mm thick (shipped in equal lengths) and strips 0.1 to 5.0 mm (shipped as slit coils, spooled or as strips). The thickness of the sheets may not exceed a tenth of the width. The semi-finished products of interest here are strips for springs and connectors standardized in EN 1654, strips for lead frames in EN 1758, plates, sheets and strips of copper for use in electrical engineering in EN 13599. The semi-finished products in these standards are listed in Table 14.

Thin flat products (foils) of Cu-ETP or Cu-FRHC are either rolled in widths normally up to 450 mm or continuously electrolytically deposited in large widths on stainless steel drums. Rolled foils have smooth surfaces on both sides while electrolytically deposited foils have a smooth and a rough surface. This is advantageous for adhesion in producing printed circuit boards for which the foil is bonded onto a synthetic support material. Electrolytically deposited foils are produced with a thickness of 10 to 100 μm and preferably 35 or 17 μm. While the price for rolled foils increases with decreasing thickness the opposite applies to electro-foils.

Reasons of rationalization have in the past led to the transition from sheets to strip, for example, for producing pressings. Strip is delivered either as coils or on spools. This means that current lengths of normal coils no longer satisfy the demands of faster running, higher performing power presses. However, in the case of greater lengths, coils that are wound layer upon layer are large and unstable. To achieve greater lengths, the strip is spooled on a relatively wide

Table 14 Dimensions of flat products of copper materials in automotive engineering.

Dimension	Strips for springs and connectors (EN 1654)	Strips for lead frames (EN 1758)	Plates, sheets and strips for use in electrical engineering (EN 13599)
Thickness [mm]	from 0.1 to 1.0	from 0.1 to 2.0	from 0.05 to 25
Width [mm]	≤350	–	from 10 to 1250

core, coil by coil and layer upon layer. Strips are either delivered wound on cardboard supporting rings or spooled on paper cases or spools.

Dimensions and weights of the coils and spools differ and are subject to agreement when ordering. The relation between strip size, coil weight and strip length is explained in the following example: With a strip size of 0.30 mm×20 mm, a strip of brass weighs 0.051 kg per 1 m length. With a specific coil weight of 4.5 kg/mm strip width, a ring weighs 90 kg with a strip length of approx. 1.8 km. If this strip size for example is spooled to a 1 tonne coil, the strip length will be approx. 20 km. Spring strips are delivered in widths up to 350 mm, whereby the widths can be slit on modern strip slitters into all required widths. Sheets and strips can be cut into required widths between 500 and 1100 mm and maximum lengths of 5000 mm.

Some mills producing semi-finished goods supply contour-milled strips. These strips with profiled cross section enable the dimensional adaptation of a spring component with different local stress. These strips are produced by milling, peeling, rolling or welding and have the following advantages:

- There is no work hardening due to reducing the thickness.
- Bendability also remains in the thin areas.
- The required forming forces are less and equal over the strip width.

Strips of copper and copper alloys are supplied as hot-dip tinned strips with layers of ≥5 μm to improve soldering properties and with layers of approx. 1 μm to improve the tarnishing resistance. The hot-dip tinned strips – differentially tinned on both sides or in stripes in a longitudinal direction – with the coating 0.7 to 13.0 μm thick are standardized in EN 13148. Electrolytically tin-plated strips with coatings 0.8 to 6 μm in places are standardized in EN 14436.

4.2.2 **Tubes** [28]

Seamless circular tubes of copper and copper alloys for general use are standardized in EN 12449 and seamless copper tubes for use in electrical engineering are standardized in EN 13600. These specifications contain limiting dimensions and shape tolerances. EN 12449 applies to the range from 3 to 450 mm outer diameter and to wall thicknesses of 0.3 mm to 20 mm. EN 13600 lays down dimensions for profiles and sizes of copper tubes for electrical engineering (supplied in straight lengths) for:

- circular tubes with outside diameters from 5 to 150 mm and wall thicknesses from 0.5 to 20 mm;
- square and rectangular tubes with the largest outside dimensions from 5 to 150 mm and wall thicknesses from 0.5 to 10 mm.

The diameters available cover all conceivable dimensions in automotive engineering. For tubes it is also valid that the available dimensions depend on the strength state.

Seamless, rolled finned tubes for heat exchangers (fins improve heat transfer) are standardized in EN 12452. It applies to the dimensional range from 6 to 35 mm outside diameter and from 1 to 3 mm wall thickness in the unfinned section with a fin height up to 1.5 mm. It should be mentioned here that copper heat pipes with extremely high thermal conductivity are also commercially available.

Geometrically speaking, hollow rods can also be seen as thick-walled, extruded tubes. They are however treated here in accordance with the division implemented in the specification (see next chapter).

4.2.3 Rods

Rods of copper and copper alloys are standardized in EN 12163 (for general use) and in EN 12164 (for machining purposes); EN 13601 deals with rods (and wires) consisting of copper and copper alloys for electrical engineering. The cross sections of the rods are shaped as circles or regular polygons (e.g. squares, hexagons or octagons). Products with polygonal cross sections may have rounded edges over their entire length.

The diameter and across flats size of rods for general use, depending on material and condition, is 1.6 to 80 mm whereby the lengths and their limiting dimensions must correspond with the requirements stated in the enquiry or the order. Rods for machining can have diameters from 2 to 80 mm, depending on the material and the condition, and across flats sizes from 2 to 60 mm; preferred (available) lengths are 2000, 2500, 3000 or 4000 mm depending on diameter and across flats size. Furthermore, precision rods faced and chamfered with narrow tolerances have been developed specially for machining on automatic lathes with automatic rod feed.

Limiting dimensions and shape tolerances for copper rods for electrical engineering are laid down in EN 13601. According to this, the cross sections and dimensional ranges are:

- round, square and hexagonal rods with diameters or across flats sizes from 2 to 80 mm;
- hexagonal rods with thicknesses of 2 to 40 mm and widths of 3 to 200 mm.

Rectangular rods (besides profiles) of copper or copper alloys for general use are standardized separately in EN 12167. This specification applies to rectangular rods (delivered in equal lengths) in thicknesses of 3 to 60 mm and widths of 6 to 120 mm.

Hollow rods are a straight product with a uniform cross section and a hollow space over their entire length. The hollow rod's inner and outer contour can be round, square, rectangular, hexagonal or octagonal on any section. Hollow rods of copper and copper alloy for machining are contained in EN 12168. The outer diameters and across flats sizes can be between 12 and 80 mm, depending on material and condition, and the limiting dimensions for the bore diameters are

8 to 70 mm. They are supplied in fixed lengths whereby the lengths correspond to those of the rods for general use.

4.2.4 Wires

Copper and copper alloy wires for electrical engineering are dealt with in EN 13601 according to which the wires may have square, hexagonal or rectangular cross sections and rounded edges over the entire length. Limiting dimensions and shape tolerances of the wires are laid down in the said specification, the cross sections and dimensional ranges are:

- round, square, hexagonal and rectangular wires with diameters or across flats sizes from 2 to 25 mm and for thicknesses from 0.5 to 12 mm with widths from 1 to 200 mm.

Profiled copper wires (including profiles) for use in electrical engineering are dealt with in EN 13605, in which limiting dimensions and shape tolerances of such wires with a diameter of the circumscribed circle of max. 180 mm are laid down.

Drawn round copper wires with thicknesses of 0.04 to 5 mm for electrical conductors are standardized in EN 13602 and are intended for producing rods, insulated cables and laces. The associated limiting dimensions are dealt with in this specification. It applies to bare, tin-plated, annealed or hard drawn wires and for one or multi-lead wires but not for wires for lacquering, for use in electronics or for contact wire for electric drive operation. For tin-plated wires, the thickness of the unalloyed tin coating for types A and B is shown as 0.3 or 0.6 μm while it is not stated for type C.

The highly conductive copper drawing stock for wire drawing, mainly for producing electrical conductors, is dealt with in EN 1977. This specification deals with drawing stock in nine copper grades and nine silver-bearing copper grades. It applies to sections approximately the same level of roundness and with a diameter from 6.0 to 35 mm (limiting dimensions for larger diameters are to be agreed separately).

Copper and copper alloy wires for general use and for springs and fasteners are dealt with in EN 12166. This lays down limiting dimensions for the diameter of round wire from 0.25 to 18.0 mm, for the across flats size of square and regular polygonal wires from 0.50 to 18.0 mm and for the width and thickness of rectangular wires from 1.0 to 18.0 mm and edge radii for square or rectangular wire from 0.6 to 12.0 mm. The wires are shipped in rings or coils whereby large weights are important for economic efficiency on cutting machines. Coil weights of 1000 kg are usual, greater weights are possible by agreement with the manufacturer.

For resistor wire there are no recent national specifications, an alloy of 56% Cu + 44% Ni, known as HECNUM is available in the UK and a similar material called KONSTANTAN is available in Europe. Typically these materials can be supplied in sizes ranging from 0.1 mm to 5.5 mm diameter. They are characterized by a very small temperature coefficient of electrical resistance.

4.2.5 Drawn and Extruded Profiles

Extruded profiles are mainly made of leaded copper–zinc alloys and copper–zinc alloys with other alloying elements with various levels of strength. Copper and copper alloy profiles for general use are dealt with in EN 12167. The cross-sectional dimensions of profiles must correspond with the limiting dimensions in the drawing provided by the customer and agreed by the supplier.

The possibilities for producing profiles are heavily dependent on the shape and, for economic reasons due to the manufacture of the tools, highly dependent on the lot size. It is advisable for the designer to get in touch with the manufacturer as soon as possible to clarify the production possibilities and the economics. The smallest diameter of the circumscribed circle of a profile in general ranges from under 20 to approx. 150 mm ∅, the smallest web thickness, depending on shape and nominative diameter, for extruded profiles from 2 to 6 mm and for drawn profiles from 1.5 to 3 mm.

Profiles and profiled copper wires for use in electrical engineering are dealt with in EN 13605, in which limiting dimensions and shape tolerances of profiles with a diameter of the circumscribed circle of max. 180 mm are laid down.

4.2.6 Forgings

EN 12420 lays down dimensional and shape tolerances for drop and open die forging. The minimum wall thicknesses in hot stamping are about 2 mm, depending on the size of the forging. Floor thicknesses, draught of the mold, rib thicknesses and possible molds are contained in the specification. Also here, scope for design, material requirements such as strength, etc. and economics are to be agreed with the manufacturer as soon as possible.

Copper and copper alloy open die forgings are of no importance in automotive engineering due to their weight and owing to the necessary lot sizes.

Wrought and unwrought copper and copper alloy forging stock is standardized in EN 12165.

4.2.7 Special Shapes

There are no EN specifications for special shaped parts, such as sintered products, electroforming etc., containing shapes and dimensions. Information on sintered products is contained in some literature from the German Copper Institute [1, 24, 29] or is available from the German Powder Metallurgy Association [Fachverband Pulvermetallurgie], Goldene Pforte 1, 58093 Hagen, Germany.

4.3
Classification and Designation

EN specifications are binding within the European Union. The following general definitions apply in the application of European specifications for copper materials:

- **Copper material**
  Generic term for copper and copper alloys
- **Standardized copper material**
  Copper material standardized in a European specification
- **Non-standardized copper material**
  Copper material not standardized in a European specification but produced and/or used in Europe.

4.3.1
Designation by Material Number

The material numbering system described in EN 1412 is based on the appendix to ISO/TR 7003, in which a copper material is allocated a certain number for classification into categories. This material numbering system is an alternative to the material designation system with codes as laid down in ISO 1190-1.

According to this system, all copper material numbers must consist of six characters. According to ISO/TR 7003, the code in the first position must be the letter "C" in order to describe the copper material.

The character in the second position must be a letter identifying the material group. Table 15 lists the respective letters and their meaning.

The characters for the third, fourth and fifth positions must form numbers between 000 and 999 with these characters being used for numbering individual materials within a certain material group. The numbers 000 to 799 refer to standardized copper materials and those from 800 to 999 relate to non-standardized materials; otherwise, these codes have no specific meaning.

The character for the sixth position must be a letter identifying a material group. Table 16 shows the positions from 3 to 6 and their meanings.

Table 15 Letters and their meanings in the material numbering system (position 2).

Letter	Meaning
B	materials in ingot form (e.g. blooms) for re-melting in producing cast products
C	materials in the form of cast products
F	filler materials for welding and brazing
M	master alloys
R	refined copper in raw forms
S	materials in the form of scrap
W	wrought materials
X	non-standardized materials

Table 16 Meaning of the positions 3 to 5 for the material numbering system.

Material group	Positions 3, 4 and 5 (a number in the following area)		Position 6 (letter describing the material group)
	Standardized copper materials	Non-standardized copper materials	
Copper	000 to 799	800 to 999	**A** or **B**
Low alloyed copper alloys (Alloying elements less than 5%)			**C** or **D**
Special copper alloys (Alloying elements less than 5%)			**E** or **F**
Copper–aluminum alloys			**G**
Copper–nickel alloys			**H**
Copper–nickel–zinc alloys			**J**
Copper–tin alloys			**K**
Copper–zinc alloys, binary alloy			**L** or **M**
Copper–zinc–lead alloys			**N** or **P**
Copper–zinc alloys, multi-alloy			**R** or **S**

4.3.2 Designation According to Chemical Composition

In agreement with the findings of the Technical Committee CEN/TC 133 (in accordance with ISO 1190-1) details on the material codes are described together with the chemical symbols of the alloy components. Here, the chemical symbol "Cu" for the base metal copper is used as a prefix without showing its weight proportion followed by the chemical symbols of the main alloy components mostly in the order of falling weight proportions. The weight proportion of each element is indicated by the measured value in percent directly (without blanks) following onto the chemical symbol. In the case of several alloying elements e.g., more than two or three, the indication of the weight proportion for the other alloying elements with lower weight proportions can be deleted in individual cases.

Example of the designation of a copper alloy according to EN:
Material code: CuZn37Mn3Al2PbSi
Material number: CW713R

Material codes for the unalloyed copper or phosphorus-bearing copper types are the exception here whereby code letters separated by a hyphen are added to the prefix "Cu" to identify the production process or refer to the residual elements. They are the abbreviations for the following English terms:

ETP **E**lectrolytic **T**ough **P**itch
FRHC **F**ire **R**efined **H**igh **C**onductivity
FRTP **F**ire **R**efined **T**ough **P**itch

OF Oxygen Free
PHC Phosphorus Deoxidized High Conductivity
DXP Deoxidized Phosphorus

Example of the designation of a copper type according to EN:
Material code: Cu-ETP
Material number: CW004A

Oxygen-containing copper with at least 99.9% pure Cu and a minimum conductivity of 58 MS/m.

4.3.3 Designation of the Condition of the Material

The designations of copper and copper alloy materials' conditions are contained in EN 1173. This European standard lays down the system for denoting conditions of copper material to be used for identifying compulsory requirements for their properties. The designations contained in these apply to products of wrought or cast materials excluding ingots for foundries. Here, the manufacturing and/or heat treatment processes to achieve the requested properties were intentionally not stated but left to the decision of the manufacturer.

The designation of the condition according to EN 1173 normally consists of four codes beginning with a letter followed by three digits. In the case of additional treatment there must be another digit in the fifth position or a suffix must be added in the fifth or sixth position. Only one designation is permitted for a certain condition and for the entire dimension range for which the minimum requirement applies.

The capital letter in the first position describes the mandatory property laid down in a corresponding product specification. The letter used to identify the property does not exclude the combination of 2 or 3 properties if these are laid down in the respective product specification. The code letters according to Table 17 must be used.

Table 17 Code letters of material properties according to EN 1173 (first position).

Letter	Obligatory properties to be described
A	fracture elongation
B	spring bending limit
D	as drawn, without specified mechanical properties
G	grain size
H	hardness (Brinell or Vickers)
M	as manufactured, without specified mechanical properties
R	tensile strength
Y	0.2% proof stress

The code letter is followed by a three-digit number except for the letters D, G and M in order to identify the *minimum* values of the specified properties in the respective product specification. There are no codes following the letters D and M as there are no obligatory properties laid down for these conditions. After the letter G, which identifies the grain size, there is also a three-digit number indicating the *average* grain size as an obligatory property laid down in the product specification.

Leading "0's" are to be placed in front of the second or third digit to fill in three-digit numbers in cases where the property's measured value contains only one or two digits. The fifth position will also be used if the measured value contains four digits, for example, in the case of high tensile strengths or heat-treated alloys.

If additional heat treatment is requested for the purpose of stress-relieving of a product the suffix "S" is used in position 5 or 6.

The designation of the condition should be used in the product designation and ordering information, which are followed by the material designation and separated by a hyphen ("-").

Designation examples according to:

Tensile strength	Rods EN 12164-CuZn39Pb3[a)]-R500-.....[b)]
0.2% proof stress:	Strips EN 1654-CuZn30-Y460-.....[b)]
Elongation:	Wires EN 13602-Cu-OF-A007-.....[b)]
Drawn:	Tubes EN 13600-Cu-ETP-D-.....[b)]
As manufactured:	Hollow rods EN 12168-CuZn36Pb3-M-.....[b)]
Hardness:	Sheets EN 1652-CuZn37-H150-.....[b)]
Grain size:	Strips EN 1652-CuZn37-G020-.....[b)]
Spring bending limit:	Strips EN 1654-CuSn8-B410-.....[b)]
Additionally stress-relieved:	Rods EN 12164-CuZn39Pb3-R500S-.....[b)]

a) Material numbers can also be indicated here instead of material codes, e.g. here, CW614N.
b) Further addition according to the respective product specification e.g. nominal sizes, tolerances or design specifications.

4.3.4
Product Designation

A complete, fast and clear description of a product is demonstrated by the product designation, a standardized designation model. This model creates the condition for mutual understanding on an international level regarding products which correspond to the respective European standards. Details required for product description are stated in the respective product specifications.

The derivation of a product designation in accordance with EN 13599 (plates, sheets and strips of copper for use in electrical engineering) is explained in the following example.

Sheeting used in electrical engineering in agreement with this specification consisting of the material Cu-ETP (or. CW004A), in the condition H040, with a

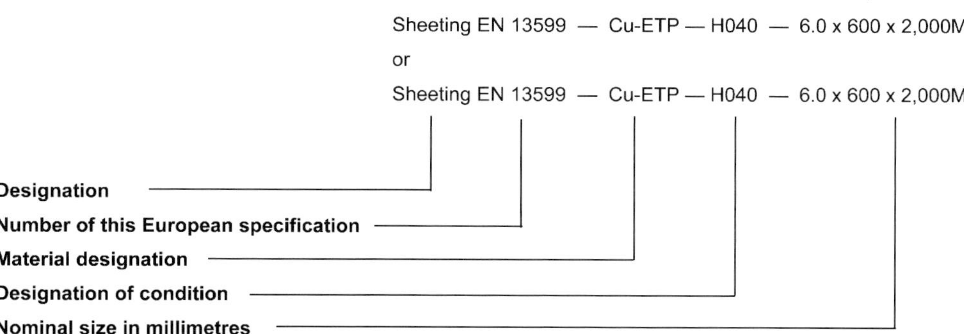

Fig. 2 Example of a product designation according to EN 13599.

thickness of 6.0 mm, a width (nominative dimension) of 600 mm and a manufacturing length of 2000 mm must be described according to Fig. 2.

4.3.5 Designation of Powder Metallurgical (Sintered) Materials

The designation of powder metallurgically produced copper materials is not standardized in EN specifications. Manufacturers of sintered products use their own company designations instead. Here, the abbreviation “SINT” often precedes the material designation. For example, SINT B/C 50/51 is used to designate various sintered bronzes based on CuSn10, as standardized in ISO 5755.

If sintered materials are required, it is advisable to contact the Powder Metallurgy Association (address see Section 4.2.7) and have manufacturers of PM materials verified.

5
Copper Casting Materials [30]

Copper and copper alloy ingots and castings are dealt with in EN 1982. Unless stated otherwise, the mechanical properties laid down in this specification refer only to separately cast test bars that serve as estimators of the product quality and casting technique. The strength established in one or several separately cast test bars may differ from that of the castings themselves and cannot be expected in all areas of the casting. This is due to possible differences in the microstructure between test bars and castings. Foundries however, have the possibility to obtain mechanical strength properties in previously selected parts of the casting. These requirements are to be agreed beforehand between the customer and the foundry. It is advisable for the designer to contact the foundry at an early stage in the design phase.

If tensile properties are set for the certificate of conformity as per EN 1982, the request must be carried out on proportional test pieces according to EN 10002-1.

The diameter of sand casting test bars must be between 12 and 25 mm; they must be separately cast in sand molds from the same melt which they represent. The test bars for chill casting are to be cast in ingot molds. If the test bars in both cases are to be machined, the actual diameter must be between 10 and 18 mm and correspond to EN 10002-1.

In centrifugal casting, test bars are cast separately or taken from the casting itself subject to determination of the customer. If the customer requires test pieces from castings of more than 50 mm thickness position and properties must be agreed in advance. If test bars are required for pressure die castings, a separately cast, flat test bar (2 to 4 mm thick) is to be used and tested without being machined.

In the case of continuous casting, test bars must be taken from the casting itself whereby the gauge length must be parallel to the casting direction.

5.1
Casting Procedures [21, 30]

Copper materials have a relatively high specific weight. As weight saving is of the utmost importance in automotive engineering, only the smallest possible castings of copper materials are used and only then if such material properties

Copper in the Automotive Industry. Hansjörg Lipowsky and Emin Arpaci

ISBN: 978-3-527-31769-1

are required which cannot be achieved with specific lighter or cheaper materials. But cost for machining and other necessary processes should also be considered.

Casting is the shortest way from stock material to the finished product. When forming by casting, the same molding and casting procedures are used for copper castings as for iron and other non-ferrous metals. Copper casting materials are basically suitable for all traditional casting procedures. For technical and economic reasons, by far the largest number of all copper materials castings are produced in sand casting, chill casting, centrifugal casting, continuous casting and to a lesser extent (brass) pressure die casting. The casting procedures, molding and core production are only roughly described in the following. More information is available in the manual "Guss aus Kupferlegierungen" ["Copper alloy castings"] [30].

In casting procedures, distinction is basically made between lost mold casting and permanent mold casting. In the first case, further distinction is made between casting procedures with permanent and lost patterns.

5.1.1 Lost Mold Casting

The following are forms of lost mold casting:

- Sand casting, hand molded with permanent patterns
- Sand casting, machine molded with permanent patterns
- Shell mold casting with permanent patterns
- Ceramic mold casting with permanent patterns
- Investment (Precision) casting with lost patterns.

5.1.1.1 Sand Casting

Sand casting is the oldest and best known casting procedure. Silica sand is usually used as molding material, either natural or synthetic molding sand bonded with clay or Bentonite. The molds are either made by hand for single or large castings or by molding machines for small and medium-size castings and large lot sizes.

All copper base casting materials can be sand cast.

Hand molded sand casting is especially suitable for producing single, normally larger, heavier castings and is therefore used, if need be, only for the manufacture of single small castings (prototypes) in automotive engineering.

5.1.1.2 Precision (Investment) Casting (Lost-wax Process)

Precision casting is characterized by the production of one-piece ceramic shell molds without mold parting lines and the inaccuracies and burrs caused by this. A wax pattern must be made for each mold and therefore every casting to be produced. The patterns are produced in dies with wax injection machines and stuck to clusters with mold parts, ingates and feeders etc. A ceramic shell is produced by immersing this several times in a ceramic slip. Sanding and drying will produce a fireproof ceramic shell. The shell is embedded in fireproof matter, the wax is melted out and the mold burned. Casting is carried out while the mold is still hot.

Precision casting is an expensive and complex casting procedure. It is suitable for all castable copper materials, particularly difficult-to-machine materials. The castings are produced without burrs and have excellent surfaces and often do not require machining – except for the fit size and the functional surfaces.

5.1.1.3 **Exact Casting (Other Processes)**

The wish to produce castings with greater accuracy and better surface qualities led to the development of some special casting processes. Examples are shell mold casting according to Croning, the gypsum mold process and the ceramic mold process. Strictly speaking, the precision casting procedure is also one of the exact casting processes.

In shell mold casting according to **Croning**, a pourable and blowable silica sand/phenol (cresol) formaldehyde synthetic resin mixture is bonded in contact with a hot metal pattern to a 4 to 8 mm thick shell. Any loose or unbonded material is removed prior to the shell being put together for casting. Highly accurate castings with a very good surface can be achieved by this procedure. It is suitable for all castable copper materials with larger lot sizes.

In the **gypsum mold process**, plaster is used as a binding agent for molding material mixtures of silica sand, fireclay, talcum, etc. The molds are either just dried or treated in a steam autoclave. Casting is then carried out.

The procedure is basically, suitable for all castable copper materials, but due to high thermal load, only for thin walled castings. It is seldom used.

The **ceramic mold process** has been known for decades as the Shaw-Form procedure. Silica sand, Sillimanite, Molochite, zirconium silicate and ethyl silicate as a binding agent are used for the molding procedure. Ceramic molds produce very accurate castings with an excellent surface quality. The procedure permits the manufacture of thin-walled castings with exact contours. High molding costs actually limit wide use, but with parts difficult to machine, the casting procedure offers an economic alternative. The procedure is suitable for all castable copper materials.

5.1.2 Permanent Mold Casting

For the production of copper and copper alloy castings permanent metal molds are used to a large extent. Greater accuracy and a better surface are achieved by the metal molds. Faster cooling and a steeper temperature gradient in the metal achieve fine grain and therefore an improvement in mechanical properties. The following are forms of permanent mold casting:

- Chill casting,

and as special processes of precision and low-pressure die casting:

- Pressure die casting
- Centrifugal casting
- Continuous casting.

5.1.2.1 **Die Casting**

Die casting and sand casting are the most commonly used casting procedures for copper and copper alloys.

The molds mainly consist of steel with high thermal shock resistance. Cast iron and copper–beryllium molds are also used. If the retractable cores are also of metal, as in precision die casting, this is called full die casting. It must however be borne in mind that certain undercuts in the mold are not possible, even with retractable cores.

With gravity die casting, the die is tilted in order to achieve a low height of fall and a calm, turbulence-free, i.e. laminar mold filling. Here, air must be able to escape from the mold. The die is set up with increasing mold filling. Mold filling must take place as fast as possible but may not exceed a certain speed.

The die casting process offers the following advantages: A finer microstructure and greater strength than by sand casting, better, smooth surfaces, accuracy and pressure tightness, minimization of machining and the possibility of recasting inserts of other metals. The only disadvantage is the higher molding cost. The low-pressure die casting procedure is used for copper materials as for other casting materials but is of no technical importance for automotive engineering due to the larger castings.

The die casting procedure is suitable for all copper casting materials with narrow solidification range. These are low and unalloyed copper, copper–zinc and copper–aluminum casting alloys.

5.1.2.2 **Pressure Die Casting**

In pressure die casting, the mold and cores consist of a hardened hot working tool steel. Mold filling takes place in a pressure die casting machine at high speed and solidification is under high pressure. As copper casting materials have relatively high melting points compared with those of other non-ferrous metals, the thermal shock load to which the mold is subjected is extremely high and therefore the service-life is relatively low. For this reason, this casting procedure has not become established for copper casting materials. It is only used to a lesser extent for the comparatively low melting copper–zinc casting alloys. Short service-lives of the molds greatly affect the economics of the process.

Of the copper casting alloys only the materials CuZn33Pb2Si-C-GP, CuZn35PbAl-C-GP, CuZn39Pb1Al-C-GP, CuZn39Pb1AlB-C-GP, CuZn32Al2-Mn2Fe1-C-GP and CuZn16Si4-C-GP are suitable for pressure die casting.

5.1.2.3 **Centrifugal Casting** [31]

Centrifugal casting, due to the inherent nature of the procedure itself, is only suitable for producing rotationally symmetric castings. In the centrifugal casting procedure, a chill mold is put into horizontal or vertical rotation and is filled with liquid metal. The wall thickness or inside diameter of the solidifying tube-shaped casting is determined by the amount of metal added. The solidification of the melted metal takes place very rapidly and can be increased by additional cooling. A dense fine-grain microstructure is achieved by the influence of centrifugal force and extremely fast solidifying. This casting procedure is excellently suited to produce plain bearings and together with continuous casting is of great importance for producing stock material for plain bearings.

Centrifugal casting is specially suited for copper–zinc, copper–tin, copper–tin–zinc and copper–aluminum casting alloys. The casting procedure is of limited suitability for copper–tin–lead casting alloys due to the possible segregation of lead under the influence of centrifugal force.

5.1.3
Continuous Casting [31]

Continuous casting is of great technical importance in the production of semi-finished products as regards certain formats such as slabs and billets. Furthermore, the continuous casting procedure is used for producing castings with the characteristics of finished parts, mainly as stock material for plain bearings, bushings and sliding rods. Regarding casting technology, continuous casting is superior to other casting procedures as there are stationary casting and solidification conditions both for continuous and semi-continuous casting.

In the continuous casting procedure, as much liquid metal is poured into a short, intensively cooled chill mold as at the same time solidifies in the mold and is extracted. The chill mold consists of either copper, low alloyed copper or graphite. The “endless” solidifying metal strand, that has only solidified at the outer shell, is further intensively cooled on extraction (secondary cooling) and cut by a flying saw into fixed lengths. Extraction is carried out in uneven steps, interrupted by brief standstill or reversal times. Distinction is made between horizontal and vertical as well as continuous and semi-continuous casting. The investment costs are greater for vertical continuous casting due to the large construction heights required. But for horizontal casting there is the danger, for thicker sections, that the hot, not yet fully solidified strand will become deformed under its own weight. At semi-continuous casting, the cast is interrupted when reaching the intended strand length and then started again, possibly with material of another chemical composition.

It is possible to cast hollow strands by mandrels fixed in the chill mold. All strands with regular cross sections, e.g. round, square or rectangular, full, hollow or flat, can be produced. There is a dense fine grain microstructure as the strand is continuously fed and solidification is rapid and strictly unidirectional, similar to centrifugal casting. The good mechanical properties that can be achieved are identical for continuously cast strands and centrifugal casting formats. The procedure is excellently suited to produce feed stock for the manufacture of plain bearings.

The continuous casting procedure is suitable for all copper casting materials. Feed stock for plain bearings is produced above all from copper–tin, copper–tin–zinc and copper–tin–lead casting alloys by continuous casting.

5.1.4 Composite Casting

The strip-pouring process for producing thin-walled wrapped bushings with infused copper–tin–lead casting alloys for combustion engines became popular in the 1930s. Here, so-called three-component bearings with an approx. 0.35 mm thick layer of lead bronze of the alloy CuSn5Pb20-C-GS were produced. A so-called sliding layer or inlet layer of white metal (lead/tin 91/9%) is added as an "overlay".

In the continually operating strip-pouring installation, the strip that later serves as a steel backing shell of low carbon steel in thicknesses between 1.1 and 3.3 mm and widths from 100 to 150 mm runs from the coil at a constant speed into an annealing and casting unit. Both edges of the strip are bent upwards to prevent the flow off of lead bronze during dousing, bright annealed and then doused with lead bronze. The composite strip is coiled again on leaving the annealing and casting unit. The strip is milled, re-rolled, ground and cut on the processing machines to fit it for the bearing or bushing to be produced.

Recently, worm gear crowns of copper–tin alloys have also been infused onto hubs of grey cast iron or steel in the centrifugal, shell mold or die casting procedure.

5.2 Shapes and Dimensions

The castings produced are subject to various limitations in size and weight, depending on the process used. Essentially, only small and light castings can be considered for use in automotive engineering. As regards casting processes, die and precision casting and in a few cases, pressure die casting, for producing plain bearings, and the strip-pouring process for three-component plain bearings, along with continuous and centrifugal casting are common. Suitable lot sizes are required for wrapped bushings produced with very thin walls (approx. 1.5 to 3.5 mm thick) – there are no dimension limits.

Table 18 shows the approx. weight and lot size ranges for the various casting processes.

The minimum wall thickness for pressure die casting is approx. 1.5 mm, for die casting approx. 2 mm and about the same for precision casting. For centri-

Table 18 Production guidelines for various casting processes for copper alloy castings.

Mold type Type of pattern Process	Lost molds Permanent patterns				Lost patterns
	Sand casting		Shell molds	Ceramic molds	Precision casting
	Hand molded	Machine molded			
Weight range (up to approx.)	50 t	500 kg	50 kg	30 kg	1 g–2 kg (5 kg)
Quantity range (min pieces approx.)	1	20	1000	(1), 10	50 (200)

Mold type Type of pattern Process	Permanent molds without pattern			
	Die casting	Pressure die casting	Centrifugal casting	Continuous casting
Weight range (approx.)	5 g to 30 kg	5 g to 1 kg	3000 (5000) kg	30 000 kg
Quantity range (min pieces approx.)	250	3000	1	1 to 5 t

fugal casting, the existing chill mold determines the minimum quantity and for horizontal continuous casting, the minimum amount per alloy and dimension is approx. 1 tonne. However, it must be considered that manufacturers of sliding materials maintain large warehouses of various dimensions. This means that small quantities of various bushings and rod dimensions of commercially available sliding materials can be acquired ex stock.

Table 19 shows the guidelines for centrifugal and continuous casting dimensions for plain bearing production.

5.3 Classification and Designation

Due to oxidation and other processes during melting, the composition of the stock material (ingot) is changed so that the composition of the castings differs from the composition of the ingot. EN 1982 therefore generally contains the chemical compositions for the castings and the ingots.

The chemical compositions of the castings and the ingots are in many cases very similar. For some alloys, the compositions of the ingots have not been es-

Table 19 Guidelines for continuous and centrifugal casting dimensions (rods and bushings for plain bearings) for producing castings with the characteristics of finished parts.

Casting processes	**Continuous casting, horizontal**		
Formats	**Dimension ranges**		
Round rods	Outer diameter d_a [mm]		10–400
Tubes and cylinders	Outside diameter d_a [mm]		20–300
	Smallest inside diameter d_i [mm]	Small tubes	8
		Larger tubes	18
	Smallest producible wall thickness [mm]		approx. 4, however 8–10% of d_a
Square rods	Side length [mm]		20–300
Rectangular rods, plates, strips, profiles	Width [mm]		20–650
	Thickness, at least 3% of the width [mm]		≥5
Casting processes	**Centrifugal casting, horizontal rotation**		
Formats	**Tubes and cylinders, lengths much greater than the diameter**		
Plain bearing bushings	Outside diameter d_a [mm]		50–550
	Lengths		500–2000
	Smallest inside diamter d_i [mm]	raw	10
	Smallest outer diamter d_a [mm]	pre-turned	18
	Wall thickness [mm]		7–80 (200)

tablished. Their limiting value remains up to the customer and must be stated when placing the order. The ingots consisting of Cu-C and CuCr1-C are also not laid down in EN 1982.

5.3.1 Designation by Material Number and Chemical Composition

The designation of copper and copper alloy casting materials by material number is arranged according to the material numbering system explained in Section 4.3.1 and is not repeated here.

The identification of materials in EN 1982 by material codes with chemical symbols of alloy components is initially achieved analogous to the identification system laid down in ISO 1190-1 (see Section 4.3.2). Additionally (added after the hyphen), ingots are allotted the suffix B (for Block), while castings are identified by the suffix C (for Casting).

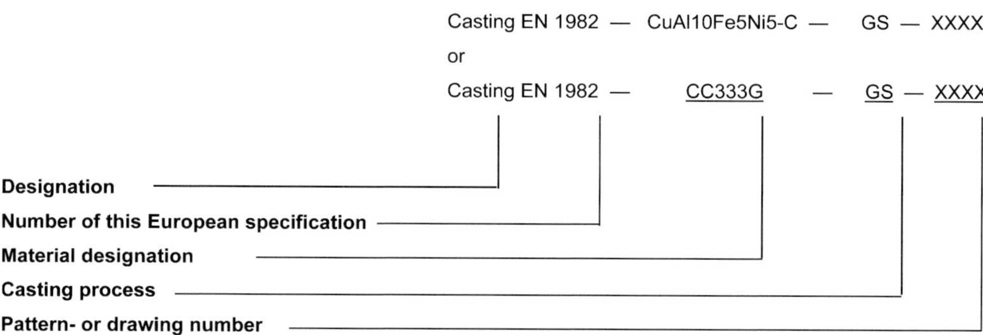

Fig. 3 Example of a product designation according to EN 1982.

5.3.2 Designation of the Casting Process and the Product

The casting processes are identified according to the designations laid down in ISO 1190-1. Here, codes relating to the casting process (see Section 3.2, Table 8) are added to the material number or the material symbol after a hyphen.

Products according to this specification must contain the following details in the product designation:

- Designation (ingot or casting)
- Number of this European specification
- Material codes or material numbers
- In case of castings, the designation of the casting process
- In case of castings, number of pattern or drawing as applicable

The derivation of a complete product designation is shown in Fig. 3 using the example of a copper–aluminum casting produced by sand casting.

6
Properties of the Copper Materials

6.1
Physical Properties [1, 14, 24]

Copper is included in the first sub-group of the periodic table, together with silver and gold. The microstructure of pure copper consists of a homogeneous phase and crystallizes in a cubic face-centred lattice. Pure copper is a salmon coloured ductile metal with a melting point of 1083 °C and a boiling point of 2595 °C. In its chemical compounds, copper is nearly always monovalent or bivalent.

Some physical properties of copper depend heavily on the purity or the contamination. On the other hand there is the possibility of changing certain physical properties over a wide range by alloying the metals and by hot and/or cold working or heat treatment. In this way, many characteristics of the copper materials can be adapted to the respective application.

Detailed information on the crystal microstructure and the physical properties are contained in the DKI [German Copper Institute] information sheets [1, 15–20], DKI material data sheets [32], material handbooks and publications [33–36]. In some cases, the product specifications contain details of an informative nature regarding some physical properties (e.g., density, Young's modulus or electrical conductivity).

For example, the product specifications EN 13599 to 13602, 13604 and 13605 contain values for specific electrical volume and mass resistance and electrical conductivity.

For an engineering metal, copper has a high specific weight. In the solid state, the *density* of pure copper at 20 °C is 8.93 g/cm^3. The density of the copper alloys changes depending on the type and amount of alloy(s) added. The density declines with the lighter elements silicon or aluminum, to a lesser extent in the case of tin or zinc and hardly changes with nickel.

The volume of materials expands or shrinks due to changes in temperature. The change in length is calculated with the *linear thermal expansion coefficient*. The volumetric expansion coefficient is about three times as large as the linear expansion coefficient. It must be observed that the coefficient of expansion itself depends on the temperature. The addition of nickel reduces the expansion coefficient of pure copper, whereas aluminum hardly influences it and zinc and tin increase it.

Copper in the Automotive Industry. Hansjörg Lipowsky and Emin Arpaci

ISBN: 978-3-527-31769-1

Table 20 Physical properties of pure copper at 20 °C.

Physical property	**Measured value**	**Units**
Lattice constant	0.3608	nm
Density (highly pure copper at 99.999% Cu)	8.959	g/cm^3
Density (technically pure copper)	8.93	g/cm^3
Density at 1083 °C, solid	8.32	g/cm^3
Density at 1083 °C, liquid	7.99	g/cm^3
Volume defect on solidifying	4.9	%
Linear coefficient of thermal expansion at RT	16.8	$10^{-6}/K$
Linear coefficient of thermal expansion from 0 to 300 °C	17.6	$10^{-6}/K$
Linear coefficient of thermal expansion from 0 to 1000 °C	20.3	$10^{-6}/K$
Melting point	1083	°C
Fusion heat	214	J/g
Evaporation heat	4770	J/g
Boiling point	2595	°C
Specific heat at 0 to 100 °C	0.38	$J/g \cdot K$
Heat conductivity at RT	>385	$W/m \cdot K$
Electrical conductivity at RT	59.62	MS/m
Specific electrical resistance at RT	1.677	$\mu\Omega \cdot cm$
Temperature coefficient of electrical resistance 0–40 °C	0.00430	1/K
Surface tension 1083 °C liquid	1.185	N/m
Dynamic viscosity at 1083 °C	4.0	$mN \cdot s/m^2$
Specific magnetic susceptibility	-0.086×10^{-6}	
Young's modulus for wrought material (upper limit for work-hardened material)	100 000–130 000	N/mm^2
Shear modulus of wrought material	40 000–50 000	N/mm^2
Poisson's ratio	0.35	
Specific electrical resistance at RT (Cu-ETP, -FRHC, -OF)	1.724	$\mu\Omega \cdot cm$
Specific electrical resistance at 300 °C (Cu-ETP, -FRHC, -OF)	4.2	$\mu\Omega \cdot cm$
Heat conductivity at RT (Cu-ETP, -FRHC, -OF)	394	$W/m \cdot K$
Heat conductivity at 300 °C (Cu-ETP, -FRHC, -OF)	377	$W/m \cdot K$

The most important and useful physical properties of pure copper are its high *thermal* and *electrical conductivity*. After silver, copper has the highest electrical conductivity of all metals. The electrical conductivity of metals is normally expressed in % I.A.C.S (International Annealed Copper Standard) or in MS/m. According to this system, 100% I.A.C.S. corresponds to 58.00 MS/m (=58.00 m/ $\Omega \cdot mm^2$). Even slight contaminations seriously reduce the electrical conductivity of pure copper. Conductivity depends on temperature. The influence on unalloyed copper is greater than for copper alloys. Work hardening reduces the conductivity of copper materials and all other metals. On the other hand, age-hard-

Table 21 Physical properties of selected wrought copper materials at 20 °C.

Material designation	Material number	Density (g/cm^3)	Melting range (°C)	Electrical conductivity (MS/m)	Thermal conductivity (W/mK)	Linear coefficient of expansion (10^{-6}/K)[a]	Young's modulus (kN/mm^2)
Cu-ETP/ Cu-FRHC	CW004A CW005A	8.9	1083	≥58.0	≥393	17.3	130
Cu-DHP	CW024A	8.9	1083	41–52	293–364	17.3	132
CuZn30	CW505L	8.5	910–965	16.3	126	19.7	114
CuZn37	CW508L	8.4	902–920	15.5	121	20.2	110
CuZn36Pb3	CW603N	8.5	885–900	14.7	100	20.6	102
CuZn39Pb3	CW614N	8.5	880–895	14.6	113	21.4	96
CuZn31Si	CW708R	8.4	880–915	8.9	71	19.2	108
CuZn38Mn1Al	CW716R	8.3	860–910	7.8	80	21.1	93
CuZn37Mn3 Al2PbSi	CW713R	8.1	875–910	7.8	63	20.4	93
CuSn4	CW450K	8.9	960–1060	12.0	90	18.2	110
CuSn6	CW452K	8.8	910–1040	9.0	75	18.5	115
CuSn8	CW453K	8.8	875–1025	7.5	67	18.5	115
CuNi18Zn20	CW409J	8.7	1025–1100	3.3	33	17.7	135
CuNi18Zn27	CW410J	8.7	1040–1090	3.3	32	17.7	135
CuNi9Sn2	CW351H	8.9	1060–1130	6.4	48	15.9	140
CuNi10Fe1Mn	CW352H	8.9	1100–1145	5.3	46	17.0	130
CuAg0,10	CW013A	8.9	≈1082	55–57	≈385	17.0	126.5
CuFe2P	CW107C	8.8	1084–1089	35	262	15.9	123
CuTeP	CW118C	8.9	1051–1081	≈54.5	≈368	18.0	120
CuZn0,5	CW119C	8.9	≈1081	≥52	≈385	17.7	125
CuBe2	CW101C	8.3	865–980	8–18	92–125	17.0	135
CuCo2Be	CW104C	8.8	1030–1070	25–32	192–239	18.0	138
CuNi2Si	CW111C	8.8	1040–1060	10–23	67–120	16.0	143
CuCr1Zr	CW106C	8.9	1070–1080	≤43	310–330	17.0	110–130
CuZr	CW120C	8.9	1073–1080	26–54	167–330	17.0	135

a) Linear coefficient of thermal expansion between 20 and 200 °C.

enable copper alloys have higher conductivity when hardened than when solution annealed. The *thermal conductivity* of pure copper is reduced slightly with increasing temperature while it increases in copper alloys.

Table 20 shows an overview of the physical properties of pure copper. Tables 21 and 22 also show the copper materials' *melting points* and *melting ranges*. While pure metals, therefore also copper, show a melting point, copper alloys melt in a melting interval. The suitability of copper casting materials for various

Table 22 Physical properties of selected copper casting materials at 20 °C.

Material designation	Material number	Density (g/cm^3)	Melting range (°C)	Electrical conductivity (MS/m)	Thermal conductivity (W/mK)	Linear coefficient of thermal expansion ($10^{-6}/K$) [a)]	Young's modulus (kN/mm^2)
CuCr1-C-GS	CC140C	8.9	≈1075	≈45	≈188	≈17.0	120
CuSn12Ni2-C-GS/-GZ/-GC	CC484K	8.6	830–1010	6.2	54	17.5	100
CuSn11Pb2-C-GS/-GZ/-GC	CC482K	8.7	830–1000	6.2	54	17.5	100
CuSn7Zn2Pb3-C-GS/-GM/-GZ/-GC	CC492K	8.8	860–1020	7.5	64	18.5	107
CuZn39Pb1AlB-C-GM/-GP	CC755S	8.5	890–910	10–14	65–85	18.0	110
CuZn38Al-C-GM	CC767S	8.5	900–915	14	85	20.0	108
CuZn16Si4-C-GS/-GM/-GZ/-GP	CC761S	8.6	830–900	3–5	34	18.0	100
CuAl10Fe2-C-GS/-GM/-GZ/-GC	CC331G	7.5	1040–1060	5–8	55	16.5	113
CuAl10Fe5Ni5-C-GS/-GM/-GZ/-GC	CC333G	7.6	1020–1040	4–6	60	18.0	119
CuSn10Pb10-C-GS/-GM/-GZ/-GC	CC495K	9.0	850–1000	6.0	54	18.7	79
CuSn7Pb15-C-GS/-GZ/-GC	CC496K	9.1	880–1030	7.0	63	18.8	78
CuSn5Pb20-C-GS/-GZ/-GC	CC497K	9.3	900–950	8.5	71	19.3	76

a) Linear coefficient of thermal expansion between 20 and 200 °C.

casting processes depends on the melting point (pressure die casting) or on the width of the solidification range (die casting).

6.2 Mechanical and Technological Properties

6.2.1 Strength Properties at Room Temperature [1, 15–22, 32]

Distinction has to be made between the strength properties of *work hardening* (non-age-hardenable or "hard by nature") and *age-hardenable* copper alloys. Work hardening alloys include those materials the strengths of which can only be increased by cold working. Increases in the strength of age-hardenable alloys can also be obtained by suitable heat treatment (precipitation hardening, age hardening) besides cold working. The strength states of semi-finished products have already been treated in Section 4.3.3. The previously exclusively used terms "soft", "semi-hard", "hard", "cold-hammered" etc. are sometimes still in use. It should also be noted that there is no sufficiently correct connection for copper materials between the easiest to determine *HB hardness* or *DPH* and the tensile strength R_m. The approximation formula:

$$R_m \approx \text{HB} \times 3.5 \text{ to } 4.0$$

is not accurate enough to calculate the strength of a copper material. Copper materials of the same composition and tensile strength can also show different values for *elongation* (A) and *0.2% proof stress* (Y or $R_{p\,0.2}$) if the materials were cold worked or heat treated in different ways during production.

Nearly all copper materials show in the stress–strain curve at room temperature a smooth transition from the elastic to the plastic range. This is why sometimes besides the tensile strength, the proof stress at 0.2% plastic deformation $R_{p0.2}$ in N/mm^2 and the fracture elongation $A5$ in % and normally also the Brinell hardness HB (or diamond penetrator hardness, DPH) are given to describe the strength properties completely. Discontinuous yielding, as for soft steels, occurs with some materials under certain conditions, e.g. with copper–zinc alloys in a soft state.

A strength increase in the case of copper alloys can be achieved by the following measures:

- Integration of alloy atoms into the mixed crystal
- Formation of a second alloy phase
- Formation of fine dispersed precipitates (including dispersion hardening)
- Influencing the grain size of the copper material

The metals Zn, Ni, Sn and Al are used as the main alloying elements for copper. The greatest increases in tensile strength are achieved with aluminum and tin. The strength increasing effect of a second phase is used for example in $(\alpha+\beta)$ brass. Fine dispersed precipitates serve to increase the strength of age-hardenable materials by heat treatment. The mechanical strength of copper materials can also be increased by the production of a fine-grain microstructure.

Copper materials are seldom used for load-bearing parts in automotive engineering. Further demands besides strength requirements [37–39] are also placed on springs, connectors, lead frames, etc. Young's modulus E can be used to assess the elastic properties. Young's modulus is the relationship between tensile stress and elastic strain in the elastic region of the stress–strain curve. Young's modulus is reduced by alloying elements such as tin or aluminum and, strongly, by the addition of zinc. With the addition of nickel however, Young's modulus rises; on addition of 30% Ni it increases from the value of 130 kN/mm^2 for pure copper to 155 kN/mm^2. Young's modulus is also reduced by cold working – except for pure copper – and by increasing the temperature. In strips that are often used for the production of elastic elements, Young's modulus can assume different values in varying directions because of textures. The lowest values are observed at 45° to the rolling direction. The loss of elastic pre-stress under load, the so-called stress relaxation is closely related to creep properties. This will be treated in more detail in Section 6.2.2.

To assess the properties of spring strips, which are treated in detail in Section 10.2, the spring bending limit *B* (also shown by FB or RFB) is an important characteristic. This is seen as a bending stress leading to a permanent deflection of 0.05 mm after relaxing the spring. This means a spring strip shows a high spring bending limit if a relatively large amount of stress can be applied with only limited plastic deformation. There is no clear mathematical relation between the values obtained in the tensile test, such as tensile strength or 0.2 proof stress, and spring bending limit. The spring bending limit is comparable with $R_{p0.1}$ (plastic deformation of 0.1%) but is considerably lower.

For bearing materials and other compression-loaded components, *compression strength* and *0.2% compression yield strength* instead of tensile strength and 0.2% proof stress are of technical importance. The characteristics determined in the compression test are generally higher than those of the tensile test.

6.2.1.1 **Rolled Products** [25, 26, 39]

The strength properties of strips and sheets are standardized in EN 1652. Rolled products of copper and copper alloys in automotive engineering are used almost exclusively as strips e.g. for connectors, lead frames, springs, etc. The values of tensile strength in the above specification range for non-age-hardenable materials from 200 N/mm^2 for pure copper up to 740 N/mm^2 for the material CuSn8. The associated 0.2% proof stresses range from values possibly lower than 100 N/mm^2 (with an elongation of 42%) to at least 700 N/mm^2. Intermediate mechanical properties are possible depending on thickness and material. The age-hardenable alloy CuBe2 according to EN shows a min. tensile strength of 1200 N/mm^2 and a minimum 0.2% proof stress of 980 N/mm^2 (for thicknesses of 1 to 15 mm) when solution treated, cold rolled and age-hardened. With this material however, tensile strengths of 1310 to 1480 N/mm^2 and 0.2% proof stresses of 1100 to 1350 N/mm^2 for thicknesses of 0.2 to 3 mm can be achieved. Strips according to EN 1652 may also be ordered with H-numbers. This means that the order is only based on hardness-numbers and that only these are to be met by the supplier. For non-age-hardenable materials, the hardness according to this specification ranges from DPH=40 of pure copper up to the highest values DPH=250 in the case of material CuNi18Zn27. In the case of age-hardenable material CuBe2 according to EN 1652, a maximum DPH of 420 is achieved.

Besides a number of other properties such as dynamic load-bearing capacity and relaxation, along with strength, it is above all electrical conductivity that is the criterion for the selection of spring strips of wrought copper alloys. Generally, wrought copper alloys for applications in electrical engineering and electro-

Table 23 Assessment of copper materials for electrically conducting springs.

Electrical conductivity (MS/m)	Assessment	Tensile strength (N/mm^2)
12 to 36	Average	400 to 600
over 36 to 48	high	over 600 to 1000
over 48	very high	over 1000

nics can be divided into three groups. This non-binding classification is shown in Table 23.

The first group was formed by materials of medium electrical conductivity with values ≥12 MS/m or 20.7% IACS. This group includes a number of copper–tin and copper–zinc alloys and the material CuSn3Zn9. The higher conductivity group with values ≥36 MS/m or 62.1% IACS includes for example the copper alloy CuFe2P. The third group was formed by materials with very high conductivity values >48 MS/m or 82.8% IACS but low strength properties such as CuZn0.5. While the materials in the first two groups are mainly used as spring materials, materials in the last group are used as lead frames for semiconductors. Spring materials are mainly used in the hard or "spring hard" state. The demands on spring strips of wrought copper alloys are standardized in EN 1654. The letters used to describe the material properties have already been compiled in Section 4.3.3 (Table 17).

Tables 24 to 27 show the values of the mechanical properties according to EN for a number of selected materials.

Mechanical properties of plates, sheets and strips of pure and silver-bearing copper for applications in electrical engineering are standardized in EN 13599.

6.2.1.2 **Extruded Products**

Extruded products of copper and copper alloys in the as-extruded state are hardly ever used in automotive engineering. Extruded semi-finished products are however the stock material for drawn products, e.g. for rods, profiles, etc., and forgings. For example, billets are often cut to length off extruded rods for the production of hot stampings. The strength properties of profiles and rectangular rods of copper and copper alloys are standardized in EN 12167. This specification contains among others a number of wrought copper–zinc alloys, leaded or with further additions such as copper–chrome–zirconium alloys.

Here, copper–zinc–lead alloys show tensile strengths of at least 280 to 480 N/mm^2, 0.2% proof stresses of approx. 130 to 330 N/mm^2, elongations of 5 to 25% and DPH of min. 85 (HB=80) to 135 (HB=130).

Copper-zinc-based multi-alloys possess higher mechanical strength. The highest value in this group is shown by CuZn37Mn3Al2PbSi, the material used to produce synchronizer rings with $R_m \geq 590$ N/mm^2, $R_{p0.2} \approx 350$ N/mm^2, elongation $A \geq 8\%$ and a DPH of at least 135 to 170 (HB=130 to 160). The half-hard conditions were supplied in the stress relieved state.

Table 24 Properties of a selected number of spring strips of work-hardened copper alloys according to EN 1654 in a tempered state with thicknesses of t=0.1 to 1.0 mm.

EN symbol		Material number	DPH [b)]	Spring bending limit R_{FB} (N/mm²) min.	Young's modulus [c)] E (kN/mm²)	Minimum bend radius for bending edge Bending test at 90° according to ISO 7438			
						Parallel to the rolling direction for thicknesses t		Transverse to the rolling direction for thicknesses t	
						to 0.25 mm	over 0.25 mm	to 0.25 mm	over 0.25 mm
CuSn5 [a)]		CW451K	160 to 190	a)	120	d)	d)	d)	d)
CuSn6	B350	CW452K	160 to 190	350	115	0	1t	0	0
	B370		180 to 210	370		1t	2t	0	0
CuSn8	B410	CW453K	190 to 220	410	115	1t	2t	0	1t
CuSn3Zn9 [a)]		CW454K	170 to 210	a)	120	d)	d)	d)	d)
			≥200	a)		d)	d)	d)	d)
CuZn23Al3Co [a)]		CW703R	190 to 220	a)	115	d)	d)	d)	d)
			210 to 240	a)		d)	d)	d)	d)
			≥235	a)		d)	d)	d)	d)
CuNi9Sn2	B370	CW351H	160 to 190	370	140	1t	3t	0	0
	B460		180 to 210	460		3t	5t	1t	2t
CuNi18Zn20	B370	CW409J	160 to 190	370	135	0	1t	0	0
	B460		180 to 210	460		0	2t	0	0

a) In EN 1654 only *R*-, *H*- and in part *Y*-values stated.
b) Not required for acceptance except in condition H.
c) Informative, not required for acceptance.
d) Only stated for state H, whereby the minimum values stated follow H.

Table 25 (continued on page 64) Mechanical properties of a selected number of spring strips of work-hardened copper alloys according to EN 1654 in a non-tempered state with thicknesses of t=0.1 to 1.0 mm.

EN symbol		Material number	DPH	Young's modulus[a)] E (kN/mm^2)	Tensile strength[b)] R_m (N/mm^2)	0.2% Proof stress[b)] $R_{p0.2}$ (N/mm^2)	Elongation[b, c)] A_{50} (%) min.	Minimum radius for bending edge Bending test at 90° according to ISO 7438			
								Parallel to the rolling direction for thicknesses t		Transverse to the rolling direction for thicknesses t	
								to 0.25 mm	over 0.25 mm	to 0.25 mm	over 0.25 mm
CuSn5	H160	CW451K	160 to 190	120	490–580	450–540[a)]	8/10	0	1t	0	0
CuSn6	H160	CW452K	160 to 190	115	500–590	460–550[a)]	8/10	0	1t	0	0
	H180		180 to 210		560–650	530–620[a)]	5/7	1t	2t	0	0
CuSn8	H170	CW453K	170 to 200	115	540–630	470–560[a)]	13/15	0	1t	0	0
	H190		190 to 220		600–690	540–630[a)]	5/7	1t	2t	0	1t
CuSn3Zn9	H180	CW454K	180 to 210	120	580–690	≥520[a)]	–/2	2t	2t	1t	1t
	H200		≥200		≥660	≥610[a)]	–/–	–	–	–	–
CuZn15	H150	CW502L	150 to 180	120	480–560	450–530[a)]	–/2	1t	3t	0	0
CuZn30	H150	CW505L	150 to 180	115	480–560	460–540[a)]	4/6	1t	2t	0	0
CuZn36	H150	CW507L	150 to 180	110	480–560	≥430[a)]	3/5	1t	2t	0	0
CuZn23Al3Co	H190	CW703R	190 to 220	115	660–750	≥580[a)]	8/10	0	1t	0	0
	H210		210 to 240		740–830	≥660[a)]	–/3	1.5t	2t	1t	2t
	H235		≥235		≥820	≥780[a)]	–/2	–	–	–	–
CuNi9Sn2	H160	CW351H	160 to 190	140	500–580	≥450[a)]	–/3	1t	3t	0	0
	H180		180 to 210		560–650	≥520[a)]	–/2	3t	5t	1t	2t

Table 25 (continued)

EN symbol		Material number	DPH	Young's modulus[a)] E (kN/mm²)	Tensile strength[b)] R_m (N/mm²)	0.2% Proof stress[b)] $R_{p0.2}$ (N/mm²)	Elongation[b,c)] A_{50} (%) min.	Minimum radius for bending edge Bending test at 90° according to ISO 7438			
								Parallel to the rolling direction for thicknesses t		Transverse to the rolling direction for thicknesses t	
								to 0.25 mm	over 0.25 mm	to 0.25 mm	over 0.25 mm
CuNi18Zn20	H160	CW409J	160 to 198	135	500–590	≥410[a)]	3/5	0	1t	0	0
	H180		180 to 210		580–670	≥510[a)]	–/2	0	2t	0	0
CuFe2P	H130	CW107C	130 to 150	125	420–480	≥380[a)]	–/3	1t	2t	1t	2t
	H140		≥140		≥470	≥440[a)]	–/–	–	–	–	–

a) Informative, respective values not required for acceptance.
b) Not required for acceptance.
c) Elongation values for thicknesses of 0.1 to 0.25/ over 0.25 to 1.0 mm.

Table 26 Mechanical properties of a selected number of spring strips of age-hardenable copper alloys in a delivery condition [a)] according to EN 1654 of t=0.1 to 1.0 mm.

EN symbol		Material number	DPH	Young's modulus [a)] E (kN/mm^2)	Tensile strength [b)] R_m (N/mm^2)	0.2% Proof stress [b)] $R_{p0.2}$ (N/mm^2)	Elongation [c)] A_{50} (%) min.	Minimum radius for bending edge Bending test at 90° according to ISO 7438: Parallel to the rolling direction for thicknesses t		Transverse to the rolling direction for thicknesses t	
								to 0.25 mm	over 0.25 mm	to 0.25 mm	over 0.25 mm
CuBe2	H120	CW101C	120 to 190	135	510–610	410–560 [d)]	15	1t	1t	0	0
	H170		170 to 220		580–690	510–660 [d)]	8	2t	2t	1t	1t
	H220		220 to 290		680–830	620–800 [d)]	2	3t	3t	1.5t	1.5t
CuCoBe2	H060	CW104C	60 to 130	140	240–380	130–320 [d)]	20	0	0	0	0
	H140		140 to 180		480–590	370–560 [d)]	2	0.6t	0.6t	0.5t	0.5t

a) Delivery condition before forming the spring (solution annealed and cold rolled).
b) Informative, details refer to the hardened condition and not required for acceptance.
c) Not required for acceptance, except for states R or Y.
d) Informative, respective values not required for acceptance.

Table 27 Mechanical properties of a selected number of spring strips of age-hardenable copper alloys according to EN 1654 in the final state[a)] with thicknesses t=0.1 to 1.0 mm.

EN symbol		Material number	DPH	Spring bending limit R_{FB} (N/mm^2)	Tensile strength[b)] R_m (N/mm^2)	0.2%-proof stress[b)] $R_{p0.2}$ (N/mm^2)	Minimum radius for bending edge Bending test at 90° according to ISO 7438			
							Parallel to the rolling direction for thicknesses t		Transverse to the rolling direction for thicknesses t	
				min			to 0.25 mm	over 0.25 mm	to 0.25 mm	over 0.25 mm
CuBe2	B530	CW101C	250 to 310	530	820–930	650–870[c)]	1.3t	1.3t	1.3t	1.3t
	B760		310 to 400	760	1060–1250	930–1180[c)]	4t	4t	3t	3t
	B820		360 to 430	820	1190–1420	1020–1280[c)]	–	–	–	–
	B880		370 to 440	880	1270–1490	1100–1350[c)]	–	–	–	–
	B920		380 to 450	920	1310–1520	1130–1420[c)]	–	–	–	–
CuCoBe2	B500	CW104C	200 to 290	500	750–940	650–900[c)]	2t	2t	2t	2t
	B590		210 to 290	590	820–1040	750–970[c)]	4t	4t	3t	3t

a) Final state achieved by the processor after shaping by hardening according to manufacturer's instructions.
b) Guidelines, not required for acceptances.
c) Informative, respective values not required for acceptance.

The age-hardenable material copper–chrome–zirconium, a high-strength copper material with high conductivity (43–44 MS/m), is standardized in the conditions "solution annealed", "solution annealed and age-hardened" and "solution annealed, cold formed and age-hardened". Tensile strengths vary according to state from min. 200 to 420 N/mm^2, the 0.2% proof stresses from approx. 60 to 350 N/mm^2, elongations from approx. 8 to 30% and the DPH from 70 to 125.

For applications in electrical engineering, the mechanical strength properties for copper and silver-alloyed copper are laid down in EN 13601.

6.2.1.3 **Drawn Products**

6.2.1.3.1 **Rods**

The details on rods of copper and copper alloys are contained in EN 12163, EN 12164, EN 12167 and EN 12168. Rods for electrical engineering are standardized in EN 13601.

In the specifications, various minimum or maximum values are standardized for most work-hardenable materials. Rods of brass, a metal that is prone to stress corrosion cracking, may only be used in the stress-relieved condition if a corrosive medium, as for instance ammonia in humid agricultural atmospheres, cannot be safely excluded and the component may be under residual or external tensile stress. Products in the M, R... or H... state can be treated specially (i.e. mechanically or thermally), in order to reduce residual stress and improve resistance against stress corrosion cracking. The suffix S is used to denote a product in the stress-relieved state.

For age-hardenable materials, mechanical properties of the rods according to materials are laid down for the conditions "solution annealed" to "solution annealed, cold formed and age-hardened".

The minimum values of tensile strengths, e.g. for rods of copper or copper alloys according to EN 12163, range from approx. 200 N/mm^2 for pure unalloyed copper, e.g. Cu-DHP, to 1300 N/mm^2 for the age-hardened alloy CuBe2.

More strength properties are laid down in the above EN specifications.

6.2.1.3.2 **Tubes**

The mechanical properties of tubes are similar to those of rods as they are both drawn products standardized in EN 12449. Corresponding mechanical properties of finned tubes for the materials Cu-DHP, CuNi10Fe1Mn, CuNi30Mn1Fe, CuZn20Al2As and CuZn28Sn1As are laid down in EN 12452.

Some automobile manufacturers use brake lining tubes (outside diameter up to 10 mm) of CuNi10Fe1Mn. This alloy shows tensile strengths of min. 290 to 480 N/mm^2, 0.2% proof stress of min. 90 to 400 N/mm^2 and elongations of min. 8 to 30% according to EN 12449.

Copper tubes are used as fuel pipes in vehicles using compressed gas as a fuel. These pipes consisting of the material Cu-DHP R290 are subject to the technical rules for installations requiring monitoring. VdTÜV material sheet 410 contains the strength requirements and calculation data for RT along with data for 100 °C and 150 °C for tubes permitted for use in gas installations in the temperature range –269 °C to 150 °C.

6.2.1.3.3 **Wires** [38]

The properties of copper and copper alloy wires in various strength levels for electrotechnical applications, e.g. wires for wire harnesses, annealed wires for electric motors, alternators, transformers, etc. are standardized in EN 13601 and EN 13602. Besides strength properties, the specification also contains details on the electrical properties and the coefficients of thermal expansion.

EN 12166 lays down the mechanical properties of wires for general use. They also stipulate the values of the mechanical properties for round spring wires.

As for spring strips, distinction is made between work-hardenable and age-hardenable materials for round spring wires [38]. In automotive engineering, round spring wires are mainly used for conductive or relay springs. Springs of CuZn36-R700 should not be used in vehicles due to the danger of stress corrosion cracking. Tables 28 to 30 show the mechanical properties of a selection of round wires of copper alloys according to the current EN 12166.

In the tables for the mechanical properties EN 12166 contains for information purposes only the formerly customary designations for strength "soft", "eighth-hard", "quarter hard", "half-hard", "three quarters-hard", "hard" and "spring hard" in comparison to the new strength definitions for wires of different materials (e.g. CuNi18Zn20, CuSn4 or CuSn8 etc.).

Table 28 Mechanical properties of a selected number of round spring wires of work-hardened (cold drawn) copper alloys in a tempered condition according to EN 12166.

Symbol		Material number	Diameter (mm) from	over	to	Tensile strength R_m (N/mm²) or Hardness	Young's modulus $E^{a)}$ (kN/mm²)	Shear modulus $G^{a)}$ (kN/mm²)	Electrical conductivity [a)] (MS/m)
CuSn6	R980	CW452K	0.1	–	0.5	min 980	115	42	9
	R950		–	0.5	1.5	min 950			
	R900		–	1.5	4.0	min 900			
	H245		1.5	–	4.0	min DPH 245			
CuNi18Zn20	R880	CW409J	0.1	–	0.5	min 880	135	45	3
	R830		–	0.5	1.5	min 830			
	R800		–	1.5	4.0	min 800			
	H235		1.5	–	4.0	min DPH 235			

a) Informative, not required for acceptance.

Table 29 Mechanical properties of a selected number of round spring wires of age-hardenable copper alloys in delivery condition according to EN 12166.

Symbol		Material number	Diameter (mm) from	over	to	Tensile strength R_m (N/mm²) min	max	Young's modulus E [a] (kN/mm²)	Shear modulus G [a] (kN/mm²)	Electrical conductivity [a] (MS/m)
CuBe2	R390 [b]	CW101C	0.2	–	1.0	390	540	135	–	9
	R410 [b]		–	1.0	10.0	410	540			
	R510 [c]		1.0	–	10.0	510	610			
	R580 [c]		1.0	–	10.0	580	690			
	R750 [c]		0.2	–	1.0	750	1140			8
	R750 [c]		–	1.0	10.0	750	1140			
CuCo2Be	R240 [b]	CW104C	1.0	–	10.0	240	380	140	–	11
	R440 [c]		1.0	–	10.0	440	560			12

a) Informative, not required for acceptance.
b) Solution annealed.
c) Solution annealed, cold formed.

Table 30 Mechanical properties of a selected number of round spring wires of age-hardenable copper alloys in the hardened condition according to EN 12166.

Symbol		Material number	Diameter (mm) from	over	to	Tensile strength R_m (N/mm²) min	max	Young's modulus E [a] (kN/mm²)	Shear modulus G [a] (kN/mm²)	Electrical conductivity [a] (MS/m)
CuBe2	R1130 [b]	CW101C	0.2	–	1.0	1130	1350	135	47	13
	R1100 [b]		–	1.0	10.0	1100	1320			
	R1190 [c]		1.0	–	10.0	1190	1450			
	R1270 [c]		1.0	–	10.0	1270	1490			
	R1310 [c]		0.2	–	1.0	1310	1520			12
	R1310 [c]		–	1.0	10.0	1310	1520			
CuCo2Be	R680 [b]	CW104C	1.0	–	10.0	680	900	140	48	25
	R750 [c]		1.0	–	10.0	750	970			

a) Informative, not required for acceptance.
b) Solution annealed and precipitation-hardened.
c) Solution annealed, cold formed and precipitation-hardened.

6.2.1.4 **Forgings**

The strength properties of hot stampings are standardized in EN 12420. This specification includes the following materials for hot stampings: copper (Cu-ETP, Cu-OF); copper–zinc alloys with and without lead, copper–zinc alloys with extra additives; copper–aluminum alloys and as individual materials CuNi10-Fe1Mn, CuNi30Mn1Fe, CuCo1Ni1Be, CuCo2Be, CuCr1Zr and CuNi2Si. The last four materials are age-hardenable.

The above specification does not generally lay down any required mechanical properties. The tensile strength, 0.2% proof stress and elongation values in brackets for selected alloys are only for information; only the hardness value must correspond with the requirements. Under this condition, the minimum values of achievable tensile strengths range from 200 N/mm^2 for pure copper to the highest values for CuAl11Fe6Ni6 with 740 N/mm^2, a 0.2% proof stress of 410 N/mm^2 and an elongation of 4% (hardness DPH = 210 or HB = 200). EN 12420 contains no strength details for the material CuZn37Mn3Al2PbSi (used for synchronizer rings) (known minimum values: $R_m = 510$ N/mm^2, $R_{p0.2} = 230$ N/mm^2, $A_5 = 12\%$, HB = 140). It must however be borne in mind that test pieces are taken from the finished product. Furthermore, all mechanical properties only apply in the direction of fiber orientation.

Open die forging is rare in automotive engineering. The mechanical strength properties for selected copper materials with the above limitation are also contained in EN 12420.

6.2.1.5 Castings

The mechanical properties for copper castings are contained in EN 1982. Although the mechanical properties of copper castings are only in the rarest of cases a selecting criterion for automobile manufacturers, Fig. 4 shows the range of mechanical properties for the various copper casting materials.

6.2.1.6 Composites

Dispersion-hardened copper with aluminum oxide – under the trade name "Glidcop" – is not standardized. According to information from the manufacturer, the material is delivered in three types with different strength properties differentiated in their quantity of aluminum oxide. The mechanical properties also vary depending on the production process and the dimensions of the semi-finished product. Table 31 shows the ranges of mechanical properties.

The mechanical properties of other composites are not standardized and enquiries should be addressed to the manufacturers.

Table 31 Mechanical properties of the dispersion-hardened copper "Glidcop" at RT.

	Tensile strength R_m (N/mm^2)	0.2% Proof stress $R_{p0.2}$ (N/mm^2)	Elongation A_5 (%)	Rockwell hardness HRC B
"Glidcop-Al-15"	345–445	225–360	16–27	55–70
"Glidcop-Al-25"	395–490	260–415	16–23	62–74
"Glidcop-Al-60"	450–535	365–440	14–23	72–90

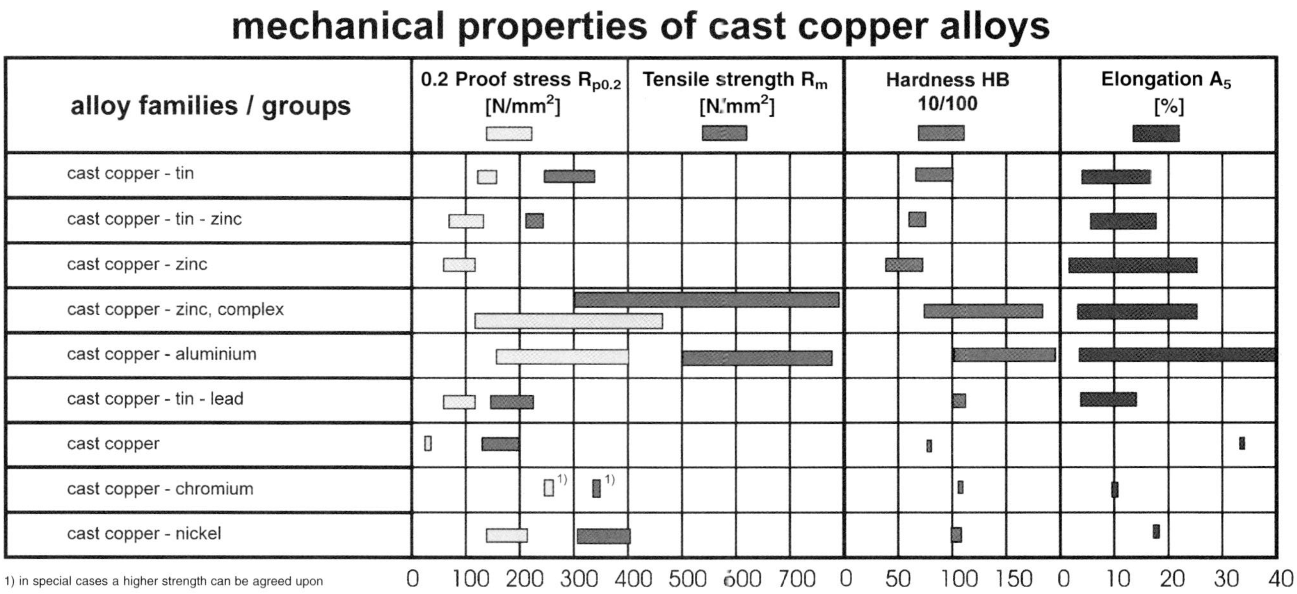

mechanical properties of cast copper alloys

alloy families / groups

0.2 Proof stress $R_{p0.2}$ [N/mm²]

Tensile strength R_m [N/mm²]

Hardness HB 10/100

Elongation A_5 [%]

cast copper - tin

cast copper - tin - zinc

cast copper - zinc

cast copper - zinc, complex

cast copper - aluminium

cast copper - tin - lead

cast copper

cast copper - chromium

cast copper - nickel

0 100 200 300 400 500 600 700 0 50 100 150 0 10 20 30 40

1) in special cases a higher strength can be agreed upon

Fig. 4 Strength properties of casting materials of copper and copper alloys [21].

6.2.1.7 **Sintered Materials**

The strength properties of sintered materials are not yet standardized in EN. They are laid down in ISO 5755 "Sintered materials – Specification". Information is available from the Powder Metallurgy Association, Goldene Pforte 1, D-58093 Hagen, Germany.

6.2.2
High Temperature Strength and Creep Properties [21, 40–44]

The use of copper materials at elevated temperatures is limited by the increasing tendency to oxidation, by the loss of mechanical strength, as determined in short-term tensile tests, and the increasing appearance of unacceptable persistent deformations or fracture at long-term loading (creep behavior) at increasing temperatures. Copper materials are generally not particularly well suited to long-term stress at higher temperatures. In this case they are used up to a maximum temperature of 300 °C.

Hot strength of the copper materials can be found in the various material data sheets [32], in the DKI brochures [1, 15–20], the specific data sheets and material data sheets from the supervisory associations and various publications. The Copper Data Sheets [43], which are unfortunately out of stock, contain the most extensive data collection of hot strength properties and also of fatigue properties. These are however available from the library of the German Copper Institute DKI.

For copper casting materials, reference is made to the DKI brochure "Guss aus Kupfer und Kupferlegierungen – Technische Richtlinien" ["Copper and copper alloy castings – Technical Guidelines"] [21], which shows high temperature strength diagrams of a number of important copper casting materials.

Table 32 shows hot strength properties for Cu-DHP-R250 and CuNi10Fe1Mn-R290 – both give examples of measured values of a high temperature tensile test.

The so-called softening temperature is also important for copper materials. It is defined as the temperature, at which after average technically normal hardening and a tempering time of 1 hour, hardness and strength start to drop considerably. For age-hardenable copper materials, the softening temperature is taken to mean the tempering resistance. This is the temperature at which the age-hardening effect starts to decline again. For long-term influence of high temperatures this is normally somewhat below age-hardening temperature. Particular attention must be paid to the softening temperature or tempering resistance in the case of subsequent heat impact, e.g. by brazing or welding.

Copper materials, like other metals, will start to creep above a limiting temperature which depends on the duration of stress. Copper materials will begin to creep, depending on the alloy group and duration of stress at approx. 150 to 200 °C. The creep properties are described by the creep strength and 0.2% creep limits. Here, the creep strength at a certain temperature is the tension required to cause a rupture in a test piece after a certain long period (e.g. 30 000 h). The

Table 32 Hot strength properties (short-term test) of materials Cu-DHP-R250 and CuNi10Fe1Mn-R290.

Temperature °C	Cu-DHP-R250			CuNi10Fe1Mn-R290		
	Tensile strength R_m (N/mm^2)	0.2% proof stress $R_{p0.2}$ (N/mm^2)	Elongation A_5 (%)	Tensile strength R_m (N/mm^2)	0.2% proof stress $R_{p0.2}$ (N/mm^2)	Elongation A_5 (%)
20	279	273	12	319	171	40
50	–	–	–	308	162	39
100	260	257	11	289	150	38
150	253	248	9	–	–	–
200	242	239	9	255	131	36
250	227	224	9	–	–	–
300	–	–	–	248	136	31
400	–	–	–	239	138	28

Table 33 High temperature oxidation resistance (scaling) of various copper materials in air.

Material group		Assessment
Designation	**Previous designation**	
Copper–zinc alloys, multi-alloys	Special brass	very good
Copper–aluminum alloys	Aluminum bronze	
Copper–beryllium alloys	Beryllium copper	
Copper–zinc alloys	Brass	good
Copper–tin alloys	Tin bronze	moderate
Copper–tin–zinc casting alloys	Red brass (lead-free)	
Copper–nickel alloys	Copper nickel	
Copper–tin–lead (–zinc) casting alloys	Red brass (leaded)	poor
Copper–tin–lead casting alloys	Copper–tin–lead bronze	
Copper, pure and low alloyed	Copper, copper-chrome	

creep strengths or creep yield limits are stated for a certain service life (long-term behavior). The indication of creep strengths and creep yield limits is set aside as copper materials are hardly ever used for load bearing parts in automotive engineering. For some materials, e.g. Cu-DHP, CuNi10Fe1Mn and others, these mechanical properties are stated in AD data sheet W6/2 and some special DKI publications [40, 41]. For some copper casting materials, some of the 0.1% creep yield limits for 10000 h in the 120 to 480 °C temperature range are known [21].

Copper materials show a sufficiently good oxidation resistance for moderately high temperatures. For applications where high temperatures but only moderate strength requirements occur, suitability is determined by the material's oxidation resistance. Table 33 enables a rough estimate of the high temperature oxidation resistance (scaling) of the alloy groups of copper. The table contains a qualitative evaluation of various copper-based material groups.

6.2.3 **Relaxation** [45–48]

Relaxation and behavior under long-term stress follow very similar physical laws. Relaxation is the subsiding of the tension of an elastically deflected spring (or other part) under stress at elevated temperature over time. Elastic strains in the test-piece are hereby transformed by certain time-dependent actions (creeping) into plastic deformations. As for creep behavior, stress relaxation properties depend heavily on material and temperature i.e. it depends on the composition of the material, the heat treatment and the grain size. The tendency to relaxation is drastically reduced by tempering in advance at 180 °C to 300 °C. Furthermore, fine-grain materials are more likely to creep and the tension more likely to drop faster than is the case with materials with normal or coarse grain.

The relaxation tendency of copper materials used for connectors is of considerable importance. In spite of their unsatisfactory relaxation behavior, critical in these applications, economic brass alloys are still used for parts with low demands on their electrical and mechanical properties. However, the superior copper–tin alloys are much preferred, but due to harsher operating conditions for connectors, even they are increasingly being replaced by materials with higher relaxation resistance e.g. CuFe2 and materials based on the CuNiSi system. Details of the relaxation properties of some copper-based spring materials are listed in the literature [45–48]. According to this, CuFe2, nickel- and silicon-bearing, low alloyed copper alloys and the age-hardenable materials CuCr1Zr and CuBe2 show a lower tendency to relaxation than copper–tin or copper–zinc alloys.

6.2.4 **Behavior at Low Temperatures**

Copper materials are excellent low temperature materials. Below room temperature, tensile strength, 0.2% proof stress and hardness and, in almost all cases, also elongation increase. In particular, the homogeneous α-copper alloys have established themselves for applications at low temperatures. In contrast to steel (except the austenitic stainless CrNi-steels), there is no steep decline in notch toughness (low temperature embrittlement) of copper materials.

The calculation and design are based on established RT mechanical properties. As nearly all mechanical properties increase with decreasing temperature, this, in practice, means an increase in the safety factor with decreasing temperature.

6.2.5 Behavior under Dynamic Load

The behavior of copper materials under high shock load as may occur during an automobile crash for example is not relevant for copper materials as copper materials are seldom used for load bearing components in automobiles. There are no results of research for corresponding dynamic stress.

6.2.6 Behavior under Cyclic Load [21, 32, 43, 49]

Under cyclic load, it is neither tensile strength nor proof stress that are important for the design of a component, but fatigue behavior or fatigue limit. Fatigue limit is the highest amplitude of alternating stress around a middle stress that a test piece can bear indefinitely without fracture. A special case of fatigue limit is oscillation amplitudes around the middle stress "0". The fatigue limit is assessed based on the "Wöhler curve". To determine the Wöhler curve, a sufficient number of test pieces are subjected to various high stress amplitudes and the number of stress cycles until fracture is determined. The easiest to determine are the fatigue properties under alternating flexural stress with the help of a flexural type test device.

For alloyed and unalloyed steel, the Wöhler curve steadily declines with increasing cycles-to-failure and reaches, unlike the copper materials – after a sufficiently large cycles-to-failure count (about 10^6 to 10^7) – a constant value, the fatigue limit. For copper materials, the curve still declines slightly at very large cycles-to-failure so, strictly speaking, there is no fatigue limit. Normally for copper materials, the fatigue limit tests are terminated at 10^7 load cycles. Therefore, the number of cycles-to-failure must always be indicated in addition to the bearable stress amplitude.

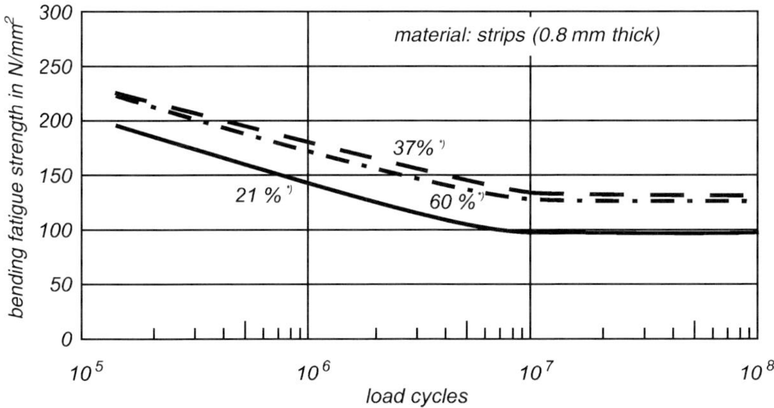

Fig. 5 Fatigue limit characteristics of strips and sheets of Cu DHP.

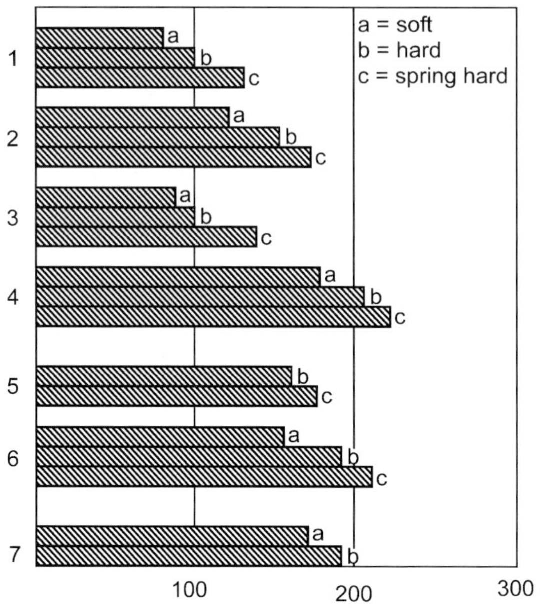

1 = Cu-DHP
2 = CuZn20
3 = CuZn37
4 = CuSn8
5 = CuNi18Zn20
6 = CuNi30Mn1Fe
7 = CuZn37Mn3Al2PbSi

Fig. 6 Fatigue properties under cyclic bending stress of a number of copper materials, sample thickness approx. 1 mm, 10^8 load cycles [14].

For the soft condition, the copper material production process or heat treatment does not affect the fatigue limit. The fatigue limit generally increases with rising tensile strength [49]. A fine grain microstructure increases the fatigue limit. Figure 5 shows the Wöhler curves for strips and sheets of various strengths of Cu-DHP.

Extensive details about strength properties under cyclic bending stress are contained in the Copper Data Sheets [43], in material data sheets [32], handbooks [14], DKI brochures and in various publications [49, 50]. For copper casting materials, current values were contained in the DKI brochure "Guss aus Kupfer und Kupferlegierungen" [21]. For bearing materials, it must be borne in mind that for composite materials (steel supporting shell with infused lead-tin-bronze), the values are to be based on the steel back. Figure 6 shows fatigue properties under cyclic bending stress for various conditions of a number of copper materials.

For shot peening to increase the fatigue limit see Section 7.4.1.7.

The influence of middle stress is taken into account by taking the bearable tension amplitudes for a given middle stress from the fatigue limit chart. The straight line of middle stress going through the zero point under 45° is plotted on this chart. For a certain material, the fatigue limit values, taken from the Wöhler curves of various middle stresses, are plotted both sides of the straight line. Approximately, the fatigue limit chart for a certain material can also be derived of the tensile strength, the 0.2% proof stress and the fatigue limit (Smith diagram).

Notches reduce the fatigue limit of a structure. Abrupt transitions, bores, grooves, etc. and also surface imperfections, heterogeneities, cast pores or corrosion damage can have the same effect. Test pieces with polished surfaces show the highest fatigue strength values. Susceptibility to notches in copper materials rises with increasing tensile strength. The designer is therefore obliged to achieve the necessary fatigue strength of a component by a more favorable design and not by increased strength of the material. As for steel, the calculations of shape factors and the resulting fatigue notch factor are to be carried out according to Neuber. The age-hardenable material CuCr1 is highly susceptible to notches. It is preferable to use the less susceptible material CuCr1Zr in its place.

Residual stress in components can reduce the fatigue strength as the residual stress adds to the flexural stress. In critical cases, it is advisable to order copper materials in the "thermally stress-relieved" state. In this condition, the material's residual stress is greatly reduced without the mechanical properties being significantly affected.

If there is additional influence of corrosion, the fatigue limit under influence of the corrosive medium is decisive. On comparison of strength properties under cyclic stress as e.g. measured in air and in seawater, copper alloys generally do better than carbon steels and also considerably better than austenitic stainless steels. There are however also considerable differences between the copper alloy groups. Copper–aluminum, copper–tin and copper–nickel alloys have proven themselves under cyclic stress under corrosive media. Applications are critical in which – even if only in traces – the appearance of ammonia cannot be excluded.

At higher temperatures, the fatigue properties of the copper materials are reduced while the values in the lower temperature range rise. A similar tendency is shown by the size of the cross section. The fatigue strength is reduced as the cross section increases.

7
Working with Copper Materials

7.1
Mechanical Processing

The mechanical processing of copper and copper alloys is divided into machining and non-cutting treatment. The explanations on processing methods are limited to the further processing of semi-finished products and castings.

7.1.1
Non-cutting Treatment [14, 26, 27]

In non-cutting forming, distinction is made between cold and hot forming. The formability of a material, i.e. the dependence of the yield strength on the previous deformation is described by the flow curve. Here, the yield strength is plotted over the deformation degree. To date there is, however, no test procedure that provides a characteristic figure for the behavior of a material both for cold and hot forming, and never will be.

Yield strength rises with increasing deformation in the case of cold forming, which takes place below the recrystallization temperature. Copper-based materials work-harden by cold forming.

On hot forming above the recrystallization temperature work-hardening is immediately extinguished by recrystallization (dynamic recrystallization). This means, the deformation degree has almost no influence on the resistance against further deformation, only the speed of deformation has.

Copper and the homogeneous copper alloys with α-mixed crystal such as α-brass (maximum 36% Zn) copper–tin and other homogeneous copper alloys have good cold forming properties. Pure copper and the material CuZn30 require the lowest energy for cold forming.

While for copper–zinc alloys the α-mixed crystal causes good cold formability, the β-mixed crystal causes good hot formability but is not easily cold formed. The heterogeneous $(\alpha+\beta)$ brass with Zn contents $\geq 37\%$ and other copper alloys, e.g. Cu–Al alloys, low and unalloyed copper – but not copper–tin alloys – are easily hot formable.

Copper in the Automotive Industry. Hansjörg Lipowsky and Emin Arpaci

ISBN: 978-3-527-31769-1

7.1.1.1 **Forming Sheets and Strips** [14, 26, 27, 38, 51]

All normal processes such as deep drawing, stretch-forming, ironing, pressing, bending, beading, chamfering or embossing, etc. are used for producing parts of sheets and strips of copper or copper alloys. Here, draw parts for automotive engineering are more of an exception. Whenever possible, it is preferable to use the excellently cold formable materials unalloyed copper and α-brass (CuZn30) for these processes. These materials are specially suited for the production of deep drawn parts.

Cold forming of age-hardenable alloys should be carried out in the solution annealed state prior to age-hardening. Cold forming in the hardened state is to be avoided.

Crack-free bending and chamfering of strips of copper alloys, e.g. for the manufacture of connectors, central electronics (electrical distributor boxes) or electrically conducting springs [51] are most important for automotive engineering. The number of interfaces and therefore connectors is rising, based on the increase in electronic components and current function expansions and the associated transition to modular design. This in turn leads to greater integration and reduction of the size of connectors. This means that requirements on the properties of copper alloy strips for the production of connectors have increased considerably. This particularly applies to bendability as connector parts have to be produced with ever-smaller bend radii following the urge to miniaturization.

The bendability of spring parts of strip for these applications is influenced by the following factors:

- Composition of the material
- Mechanical properties
- Position of the bending edge to the rolling direction (parallel or transverse)
- Geometry of the bent component (ratio of width to thickness)
- Microstructure (grain size, contamination, inclusions)

and by processing parameters for the bending procedure:

- Bending process
- Rate of deformation
- Tool surface
- Lubrication

To specify the bendability of a strip quality, the smallest usable bend radius r without cracks or the respective ratio of the radius to the strip thickness t, r/t, is indicated. In this way, the smallest possible bend radius is determined essentially by the first three factors named. The combination of bendability and strength is in many cases decisive for the assessment of a spring material.

Figures 7 to 10 show for various alloy groups the dependence of the smallest possible bend radius on the hardness and strip thickness.

The figures show that strips of CuZn37 with a hardness of 190 DPH achieve a bend radius of $r=0$ with a strip thickness of 0.18 mm. Under the same conditions, the maximum strip thickness for CuNi9Sn2 is 0.26 mm, for CuSn8 and

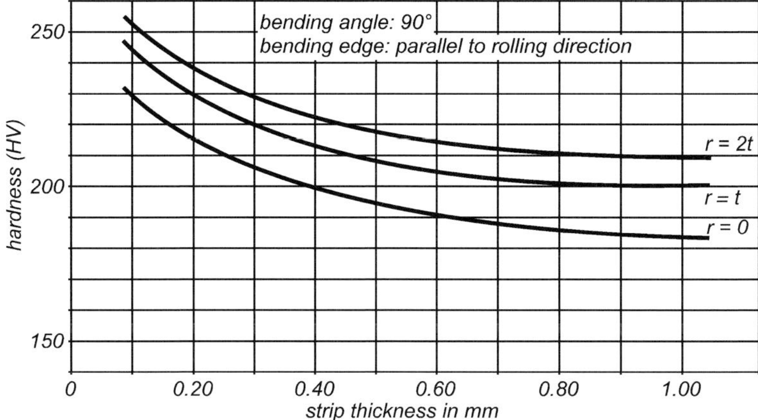

Fig. 7 Bendability of spring strips of CuSn6 and CuSn8 in a non-tempered state.

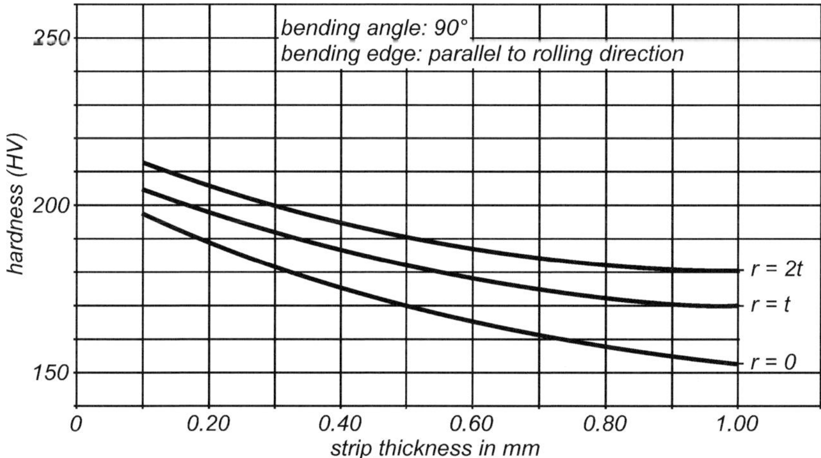

Fig. 8 Bendability of spring strips of CuZn37 in a non-tempered state.

CuSn6 they are 0.59 mm and 0.76 mm for copper–nickel–zinc alloys. Furthermore, for a strip of CuZn37 with a hardness of 190 DPH and thickness of 0.18 mm 90° bends are only possible with bend radii larger than zero.

Due to the rolling texture, besides the hardness and strip thickness, the angle to the rolling direction has a strong influence on the smallest possible bend radius. Figures 11 and 12 show that with a bending edge transverse to the direction of rolling, smaller bend radii are possible than with bending edges parallel to the direction of rolling. For technical and economical reasons however, spring parts are mostly stamped out in such a way that the following operations can be carried out most economically without any consideration for scrap and for angles to rolling direction.

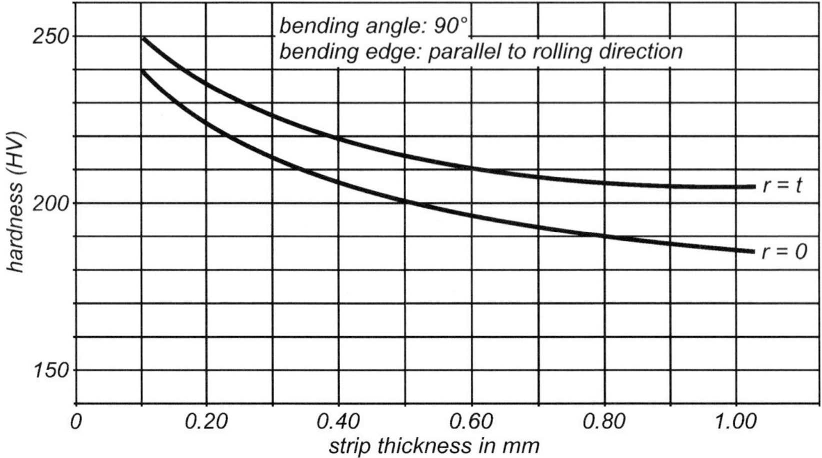

Fig. 9 Bendability of copper–nickel–zinc alloy spring strips in a non-tempered state.

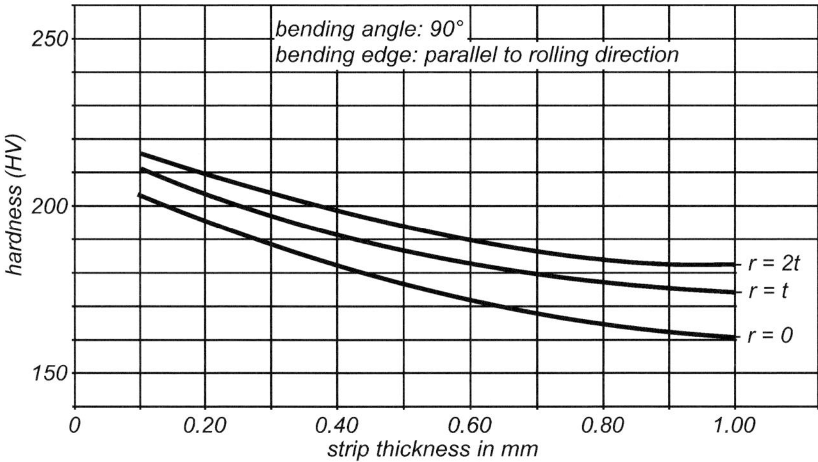

Fig. 10 Bendability of spring strips of CuNi9Sn2 in a non-tempered state.

Figures 11 and 12 show clearly that the differences in the smallest achievable bend radii of CuZn37 are relatively large and are smaller for copper–tin, copper–nickel–zinc alloys and for the material CuZn23Al3Co. This is why the latter materials are preferred for highly stressed spring parts over CuZn37. Tempering further improves bendability without strength being affected noticeably. This is particularly effective in the case of the material CuNi9Sn2, while this heat treatment is more negative for CuZn23Al3Co. The smallest radii for 90° bending for the standardized materials are laid down in EN 1654.

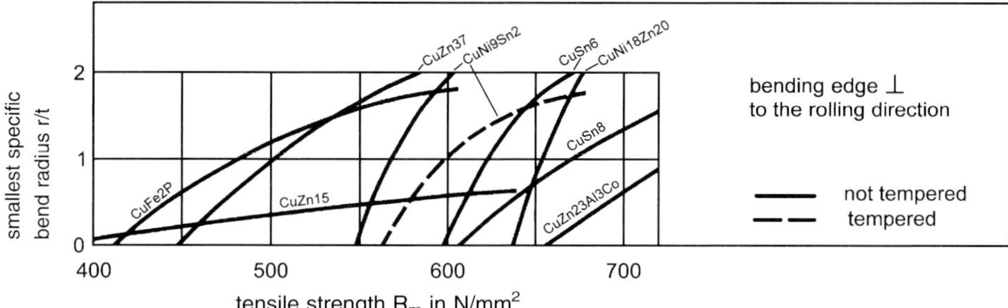

Fig. 11 Smallest bend radius in relation to strip thickness as a function of the tensile strength R_m (bending edge transverse to rolling direction).

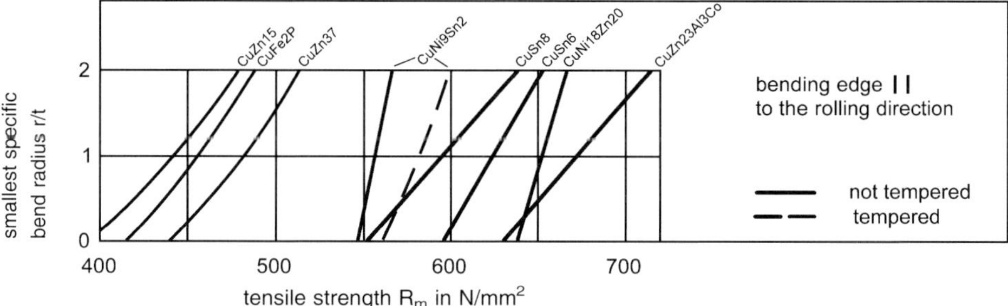

Fig. 12 Smallest bend radius in relation to strip thickness as a function of the tensile strength R_m (bending edge parallel to rolling direction).

The bend radius can be reduced further by hot bending if the achievable values are not sufficient for the intended application. This costly measure is generally not necessary for copper materials.

In recent years, the so-called contour-milled strip as stock material for the manufacture of connectors has achieved a certain market share. The increased demands, particularly for multi-pole connectors, can be better and more cost efficiently achieved with contour strips. By the use of contour-milled strips, additional stress in the components by reducing the strip thickness in the punching tool, eccentric load and changes in mechanical properties are avoided [52].

Contour strips are produced in many ways, e.g. by milling or welding. Here, reference is only made to rolling contour strips (preferred in Japan).

A number of different contour strip forms with rotation-symmetrical or non-rotation-symmetrical cut-outs can be produced. Bendability is not affected by the production of contour strips. The factor which, when multiplied by the thickness, shows the bend radius is the same for stepped and normal strips. Only in the transition range of thickness is there a reduction in bendability.

Therefore, if the material is stressed right up to the ductility limit, the bending center is to be placed outside the transition range.

7.1.1.2 **Solid Forming** [14, 53]

Solid forming can be carried out cold or hot. The production processes of cold solid forming include cold rolling, drawing, upsetting, pressing, embossing, impact extrusion, rotary swaging, thread and profile rolling. Hot solid forming comprises hot rolling, extrusion molding, hot stamping (hot pressing) and open die forging. All processes are suitable for copper-based materials, but not all copper materials are equally suitable for all processes of solid forming. Cold forming compared with hot forming requires significantly higher forces but the achievable surface qualities and the dimensional accuracy are better.

As already stated in Section 7.1.1, a rough estimate can be made of formability based on the microstructure. Of all forming processes taking place at room temperature, the homogeneous materials copper and α-brass stand out and require the least energy. For hot pressing parts (hot stampings), the heterogeneous $(\alpha+\beta)$-brass types with 60–57% Cu are the preferred materials.

Copper materials are well suited for impact extrusion, embossing, etc. But for applications in automotive engineering precision pressing and hot stamping are the main processes [54]. Pre-formed parts of leaded free-machining brass are hot stamped. Synchronizer rings for transmissions of passenger cars are hot stamped in large lot-sizes of high quality, corrosion resistant low-lead brass CuZn39Mn1AlPbSi or CuZn37Mn3Al2PbSi. In contrast, for instance copper contact members are cold (room temperature) pressed in closed dies. Materials and mechanical properties are standardized in EN 12420.

When pressing in closed dies, a flash gap is usually provided over which excess material can flow. This achieves a complete die filling without excessive energy being used in the die. Pressing and pre-forming takes place in several stages. After pressing, the product is clipped and following the last pressing, normally fettled and calibrated. Because with rising temperature in the $(\alpha+\beta)$-brass, the easily hot formable portion of the β-mixed crystal increases up to

Table 34 Forging temperatures (hot stamping) for some copper materials and flow stress k_f for different strain rates.

Material	Forging temperature T (°C)	Flow stress k_f for different strain rates	
		$\varphi^{\bullet}$ = low (N/mm^2)	$\varphi^{\bullet}$ = high (N/mm^2)
Copper, unalloyed	850–750	25–50	50–100
CuZn40Pb2	750–650	10–15	20–50
Cu–Zn-alloyed, lead-free CuZn37Mn3Al2PbSi	850–750	20–70	30–120

100%, it is important to maintain optimal forging temperatures. Table 34 shows the forging temperatures depending on the strain rate.

For hot stamping, billets of extruded or drawn stock are normally used. The material has then twice gone through an intensive plastic forming. Hot stampings have a dense and favorable microstructure which causes the favorable combination of strength, ductility and toughness. The procedure is economical for large numbers, e.g. for lot sizes of 30000 to 150000 pieces.

7.1.2
Machining [53]

In comparison to many other metallic engineering materials, the majority of copper materials are easily free-machined. The free-cutting brass CuZn39Pb3 was often used as the standard of comparison for the assessment of the materials' machinability in the production of all types of contour turned parts. Copper materials are suitable for all cutting processes. The cutting properties of various copper-based material groups however differ considerably.

There is no definition for the term "machinability" enabling the assessment of the suitability of a material for the most varied machining processes due to the complex correlations. For assessing the machining properties of copper and copper alloys, the four machinability criteria *tool wear, chip formation, cutting forces* and *surface quality* are used.

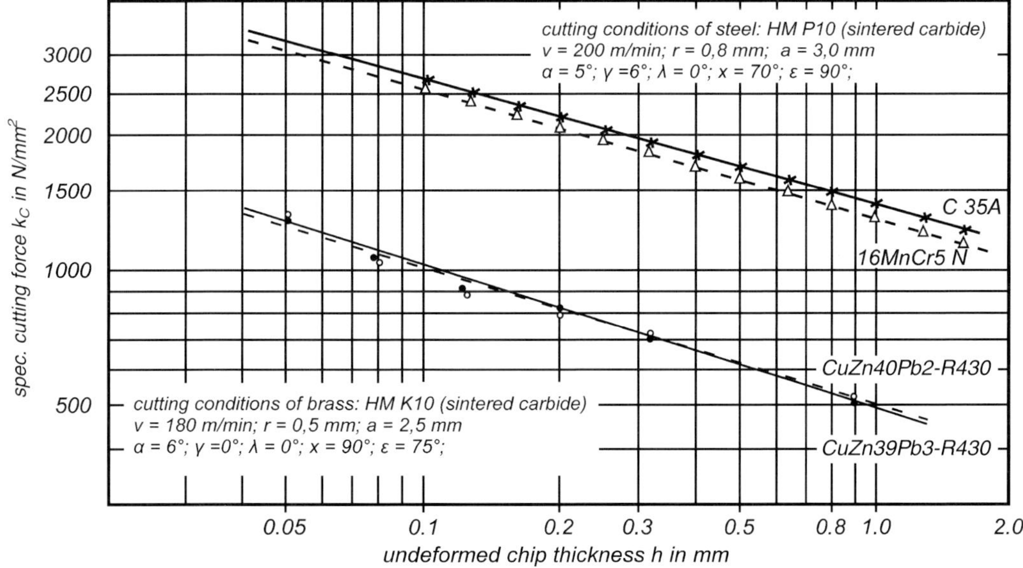

Fig. 13 Comparison of the specific cutting force of free-cutting brass with an annealed steel (C35G) and a normalized case-hardening steel (16MnCr5N) of nearly the same strength.

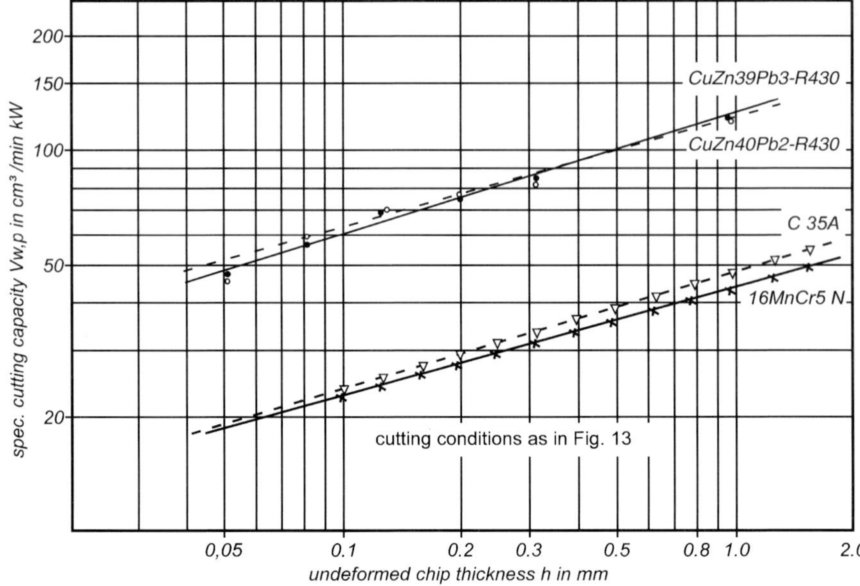

Fig. 14 Machined volume per minute and per kW power of free-cutting brass, annealed steel (C35G) and normalized case-hardening steel (16MnCr5N).

Compared with steel of the same strength, both the cutting forces and the specific machining performance of the free-machining brass alloys, i.e. the machining volumes per time and unit power, are much more cost efficient, as clearly shown by Figs. 13 and 14: If specific technical demands do not exclude its application, it is advisable to give priority to the material CuZn39Pb3 purely for machining reasons.

The formation of chips is normally the most important criterion for machinability. For this reason, a ranking of materials is generally based on chip formation. It can be generally said that the heterogeneous copper materials are more easily machined than homogeneous materials. Among the heterogeneous materials, those with a chip breaker as alloying element (normally lead) have the best machinability. However, chip breakers also considerably improve the machining properties of homogeneous materials, as e.g. those of copper by the addition of tellurium (CuTeP), lead (CuPb1P) or sulfur (CuSP). Casting alloys normally show better machinability than wrought alloys of the same composition and hardness, hardened materials machine better than soft ones. This applies in particular to homogeneous materials, as e.g. copper, unalloyed, copper–nickel, lead-free copper–zinc alloys with high copper content. Homogeneous alloys, especially in the soft condition, tend to form long thread chips that hardly break. Long chippings affect the work process, can block the machine tool or lead to the breaking of sensitive tools. The background story is explained in detail in the DKI brochure [53] that also gives guidelines for the machining of copper materials.

Table 35 Classification of copper-based material groups according to their machinability.

Main group I: Very easily machinable	**Main group II: Well to moderately machinable**	**Main group III: Moderately to poorly machinable**
Leaded copper–zinc alloys	Cu–Zn alloys (binary): CuZn36 to CuZn40	Cu–Zn-alloys (binary): CuZn5 to CuZn33
Cu–Zn-alloys with further additions leaded: CuZn40Mn1Pb1	Cu–Zn alloys (multi-alloy), all others	Cu–Zn-alloys (multi-alloys): CuZn20Al2As, CuZn28Sn1As
Cu–Zn–Ni-alloyed, leaded (nickel silver)	–	Cu–Zn–Ni-alloys, lead-free (nickel silver)
Wrought Cu alloys, low alloyed: CuSP, CuTeP, CuPb1P	–	Wrought Cu alloys, low alloyed, not age-hardenable, all others
–	Wrought Cu alloys, low alloyed age-hardenable, work-hardened	Wrought Cu alloys, low alloyed age-hardenable, hardened
Cu–Sn- and Cu–Sn–Zn casting alloys, all others	Cu–Sn- and Cu–Sn–Zn casting alloys: CuSn11Pb2-C	Wrought Cu–Sn alloys Cu–Sn casting alloys
Copper–tin–lead casting alloys	–	–
Copper–zinc casting alloys, others: CuZn33Pb2-C, CuZn39Pb1Al-C	Copper–zinc casting alloys, all others	Copper–zinc casting alloys: CuZn15As-C
–	–	Wrought copper–nickel alloys Copper–nickel casting alloys
–	–	Wrought copper–aluminum alloys Copper–aluminum casting alloys
–	–	Wrought copper unalloyed Copper casting materials unalloyed Copper casting materials low alloyed

High-speed (cutting) steel (HSS) and sintered carbides (SC) are used as cutting tools. Diamonds, often with polished facets, are used as cutting material for the highest demands on the surface e.g. the polished turning of brass. HSS is characterized by a high content of stable carbides, providing higher wear resistance and hot hardness. Steel with at least 12% tungsten is used, often also with molybdenum and cobalt. The significance of the HSS gives way in favor of the sintered carbides, especially as dry machining gains ground. As sintered carbides, the materials of the main group K, consisting of tungsten carbide and cobalt with additions of tantalum carbide and niobium carbide, are used. The widest application range is that of application group K 10, for slim wedges or weak tools, group K 15 or K 20. Exceptionally, the harder K05 is resorted to. The use of cutting ceramics or coated sintered carbides is not advised.

A rough classification of machinability in three classes has been effected depending on chip formation at turning (Table 35).

The easily machinable copper materials can be machined with high cutting speeds. Often, the performance of conventional machine tools is not enough to achieve the possible speeds. Higher performing machines must however be very rigid and have stable clamping devices.

Cooling lubricant shall be sulfur-free as the normal sulfurous lubricants used for machining steel lead to discoloration of workpieces and sometimes cause corrosion.

Detailed information and guidelines on machining of copper-based materials are contained in the brochure available from the DKI.

7.1.2.1 **Turning**

The selection of optimal cutting edge geometry and favorable cutting speed are the conditions for economic machining. Both must be adapted to the material to be processed. Guidelines for cutting edge geometry and for the cutting speeds are compiled in the DKI brochure [53]. Since all machining data depend on many, not always immediately clear, parameters and machining conditions, the guidelines can be seen as no more than benchmarks that are to be adapted, via tests with consciously varied machining parameters, to the current task. In order to avoid chatter marks with copper materials, consideration must be given to the most rigid of lathes with fixed clamping devices due to the copper materials' comparatively low Young's modulus.

Chip angle γ is especially important for the formation of the chips and their flowing off. The more a material tends to smear, e.g. unalloyed pure copper, the more peaked the wedge angle β should be. This then leads to larger clearance or chip angles. With SC tools, work is generally carried out with smaller clearance or chip angles than with HSS.

Guidelines for economic cutting speeds are more difficult to give as these are closely related to the feed rate and the chip cross section. The larger the chip cross section (scrubbing) the lower the cutting speed must be. The selection of the most favorable cutting speed also depends on what land wear is permitted, i.e., what the requirements for surface quality are.

The guidelines stated in the DKI brochure for lathing apply to a land wear of $VB \approx 0.6$ mm and a service life of $T = 30–60$ min at SC cutting of group K 10 (K 20) or $T = 45–90$ min at HSS cutting (S 10-4-3-10). When doubling the service life, the SC cutting speed is to be reduced by 16%, for HSS by approx. 10%. Exceptions are the tough, high copper-containing and unalloyed copper materials. With these, to double the service life, the cutting speed is to be reduced by

30%. If land wear is based on VB $\approx$ 0.4 mm, the cutting speed for the same service life with SC is to be reduced by 35%, in the case of HSS by 15%.

Interrupted cuts have hardly any influence on the service life of HSS, for SC, a reduction of cutting speed by 10% is recommended. When turning with obstructed chip flow (profile turning, plunge cutting chisel, parting off and thread cutting), the SC cutting speed must be reduced by about 40% and for HSS approx. 50%.

7.1.2.2 **Milling**

Copper materials are milled with all normal milling cutters, e.g. roll milling cutters, shell end mills, shank end mills, form cutters, end mills or milling heads. As far as a choice is possible, face mills are more economic than peripheral milling cutters as the blade does not rub against the processing surface in a wear-inducing manner. As long as the workpiece has no wear-inducing surfaces (e.g., scale, hardened surface, etc.) milling is normally carried out synchronously (climb cutting, up-cutting). The difference in service lives between synchronous and up-cut milling with copper materials is in fact not very great.

Due to the higher tool costs and higher operational demands on interrupted cutting, cutting speeds for milling cutters are lower than for turning chisels. As a rule of thumb, we can say that the service life of milling tools will be increased by a factor corresponding to the milling cutter's number of teeth. This applies to both SC and HSS. Since cutters with regard to tool costs and teeth numbers may be very different, guidelines for cutting speeds, see DKI brochure i. 018 [53], can only be given very conditionally. These values apply to a wear of VB $\approx$ 0.6 mm. For VB $\approx$ 0.4 mm with the same service life, the SC values are to be reduced by approx. 50% and for HSS approx. 30% (finishing cutters). It is economically mistaken to use higher cutting speeds for finishing cutting although this is recommended again and again.

7.1.2.3 **Drilling**

Copper materials are mainly drilled using HSS twist drills. There are also SC drills, reversible tips and single-lip deep hole drills.

The various machining properties of different copper materials require various cutting edge geometries or drill types for individual material groups. Materials generating short chips are drilled with drill type H, materials generating average chip shapes, with type N and extremely long-chipping materials are drilled with type W.

As long as SC drills are used, type H can be replaced by drills with brazed-in cutting plate. SC drills of type N are commercially available as so-called "K drills" with bluntly brazed sintered carbide heads.

HSS drills of type W should only be used with polished or hard chrome flanks, which is conducive to the removal of chippings. If the drillings are not too deep ($L < 2.5\ d$), SC reversing plate drills (commercially available from

$d=18$ mm ∅) are the most economical solution instead of the twist drill type N or H. This however requires powerful drilling machines. In contrast to normal twist drills, those with built-in coolant bores (from approx. $d=8$ mm ∅) are mainly of benefit for deep holes. In this way the cooling lubricant can reach the blade more easily and the liquid facilitates the removal of chippings. For extremely deep holes with, at the same time, high demands on accuracy and surface quality, SC tipped single-lip drills are normally used.

Guidelines for drilling, countersinking and reaming are contained in the DKI brochure [53].

7.1.2.4 **Cutting**

The term "cutting" refers not only to machining processes, e.g., sawing, but also chipless cutting processes, i.e. blanking, shearing and thermal cutting, to be treated later.

For copper materials, cutting takes place by sawing with hacksaws, metal band saws and cold circular saws. Cutting with cold circular saws is gaining importance. There has meanwhile been a transition from full steel blades to inserted teeth of high-speed steel or sintered carbide. Recommended chip angles are $\gamma=15^\circ$ for tough materials (e.g. copper, copper–nickel) and $\gamma=5^\circ$ and large teeth for short-chipping materials. In both cases, the clearance angle $\gamma=8^\circ$ is selected. All recommendations for cutting geometry apply to both HSS and SC. The angles are selected about 2° smaller for band saws.

For tough materials, subdividing the width of cuts has become established by alternating bevelled back teeth or by dividing the teeth in pre- and finish cutters. The first cut the middle third of the sawing gap and the latter cut the outer thirds of the slitting width. Sawing copper materials is carried out with intensive cooling by emulsion. For short-chipping materials, the emulsion is to be monitored as it can be diluted by fine flakes.

Recommendations for tooth spacing, cutting depth and speeds are contained in the DKI brochure [53].

Sheets and strips of copper and copper alloys are generally not sawn but sheared. Shearing takes place with shears, normally circular shears, or blanking dies in presses. These blanks are not just produced for further non-cutting forming but also as blanks for further machining. Blanks are above all cut from strips of the free-cutting brass materials CuZn38Pb2 and CuZn39Pb3 of which machined parts are produced for clocks and measuring equipment. Due to their high lead content, the leaded copper–zinc alloys are suitable for blanking on presses as they tend only slightly towards burr formation and built-up edges due to their high hardness.

7.1.2.5 **Fine Finishing (Polished Turning, Grinding, Engrailing, Thread Production)**

Copper materials, in comparison to other metals – with the exception of sanitary fittings – are relatively seldom ground but often fine or high lustre turned and finish milled. The most important production processes that have become established in the fine machining of copper materials are fine turning, finish milling and fine drilling; processes with which far superior surfaces, higher dimensional and shape accuracy are achieved. Also high-speed milling, e.g. with tools of polycrystalline diamond (PCD), is an example of this fine machining. These machining processes achieve surface qualities corresponding to ground surfaces. For high-speed milling with PCD-tipped blades, blade speeds of up to 1500 m/min are normal with service lives of several thousand hours. Further information on high-speed milling of copper materials is contained in a doctoral thesis [55].

Polished turning, e.g. of recording disks, reflectors, laser mirrors, etc. of copper materials, is part of micromachining and is carried out with diamond tools with defined cutting edges. Due to their extreme hardness, diamonds have a long service life and surface finishes of $R_z < 0.1$ μm and corrugations of < 0.2 μm. Often, 10 to 20 μm suffices for straightness and flatness. The main issue when turning with diamonds as tools is not increasing the cutting speed but the surface quality. Besides processing highly precise reflecting surfaces [56, 57], diamonds are also used for successfully machining commutator plates.

When grinding, the differences in behavior of the individual copper-based material groups are relatively minor. Exceptions are pure copper and the high Cu-containing materials tending to smear. For grinding, these materials require a very open structure of the working surface. Silicon carbide (SiC) is recommended as grinding medium. For peripheral grinding with ceramic binding, a grain in the region of 46–60 is selected, the hardness of J–L and the microstructure (porosity) of 4–6. The latter value for the soft high Cu-containing materials becomes 7–8, to counteract the grinding wheel being clogged. Guidelines for grinding copper materials are, as for the other processes, contained in the DKI brochure [53].

Engrailing, like thread rolling is not a machining process but a non-cutting forming. As these tasks are often carried out on lathes, they are mentioned here. In order to carry out these tasks on free-machining materials, a compromise is necessary in choosing between good machining and good cold forming properties. A good compromise is CuZn36Pb3, the main free-machining alloy in the USA. The material has such a high copper content that it shows good cold forming properties and the high lead content secures good machining properties. Details on thread rolling and engrailing are contained, e.g., in the VDI Guideline 3174.

External threads on copper or copper alloy rods on automatic lathes can be produced with a profile-cutting chisel or a chasing tool. In contrast to the production of threads with a threading die, the required accuracy can be achieved without any great effort. On automatic lathes and capstan lathes, die heads are often used that open at the end of the thread and can thus go back very quickly into the starting position. Similar conditions generally apply for the production of external threads as for turning. Cutting speeds should however be set about 40% for SC and for HSS about 50% lower than for turning.

For producing inner threads and regarding the copper materials, success depends decisively on the selection of a suitable type of screw tap. Short chipping materials such as CuZn39Pb3 or CuTeP are generally machined with straight-flute screw taps and for through holes normally a progressive cut for chipping removal. With long chipping materials such as Cu-ETP, Cu-DHP, CuZn15 etc., straight-flute screw taps with progressive cut are also preferred for up to 2 mm pitch with deep or through holes. For dead holes (blind holes) however, for right-hand thread, screw taps with a right-hand twist and angles of twist of 15°, 35° or 45° are used. The tougher the material and the larger the ratio L/d, the greater the angle of twist should be. Approximate values for cutting geometry, cutting speeds and tool service lives are indicated in the DKI brochure. As the tool costs for cutting inner threads are high, you should always check whether the material and the size of the thread do not allow production by the more cost efficient thread forming. Non-cutting processes for the production of inner threads are economically superior to screw taps due to the lack of chippings in the threads and higher revolutions.

7.1.3
Chemical Machining, Chemical and Electrochemical Polishing

The chemical machining of copper is of great technical significance in the production of "printed circuits". This means a circuit placed on a flat insulating plate of so-called circuit paths for various electronic or electrical components. The preferred stock materials for printed circuit boards are insulating plates of the most varied materials, e.g. synthetics (polyester, polyamide, etc.), or hard paper etc., laminated with a copper foil (e.g. 35 or 17 μm thick). The circuit paths are produced according to the following steps:

- Degrease the copper foil surfaces and pickle them chemically clean.
- Cover the circuit paths with acid-proof paint (normally) by the screen printing technique.
- Dissolve and remove the uncovered parts of the copper in an acid bath.
- Finally, remove the layer of paint on the circuit paths.

Etching is carried out with the most varied of mixtures of hydrochloric acid, iron(III)chloride, hydrogen peroxide, copper(II)chloride or persulfate/alkali with the addition of wetting agents.

Chemical or electrochemical processes are also used to polish copper material surfaces or to remove burrs from places that are difficult to reach, e.g. from punchings. The chemical process is the most simple but better results are achieved with the electrochemical process. Recipes for acid mixtures and temperatures to be used for chemical machining are contained in Ref. [50], p. 268.

For electrochemical fettling and polishing, the object to be fettled or polished is suspended as the anode in a special electrolyte. The microscopically small elevations and burrs are preferentially dissolved during the electrolysis and thereby the part is fettled and polished. However, not all copper materials can be brought to the same level of brilliance by this process. Lead-free copper–zinc alloys with less than 37% zinc are very suitable, the materials Cu-OF, Cu-DHP, CuZn37, CuZn40, CuZn35Pb1 or CuZn35Pb2, CuZn38Pb2, CuNi12Zn24 and CuNi18Zn20 are suitable. Electrochemical fettling can be applied to all copper materials, however copper–aluminum alloys require pretreatment to remove the layer of corrosion resistant aluminum oxide on the surface.

7.1.4 Electrical Discharge Machining

Electrical discharge machining of copper and copper alloys is hardly used in practice. Works tests have however shown that copper and brass are suitable materials for the vertical (sink, ram) eroding process. Electrical discharge wire cutting of copper is also possible [58] as shown by the results of a doctoral thesis. The suitability of a material is indeed more favorable the lower the melting point and the heat conductivity.

The use of copper materials as electrode for this process is more important. Copper and brass wires among others are used as cutting wires. For vertical eroding, electrodes of copper, copper and silver, copper and tungsten and others have become established and often produced as sintered material due to the high alloy content.

7.1.5 Thermal Cutting

As the ignition point of copper materials is not below the melting temperature, autogenous cutting, as with steel, is not possible. Copper materials would melt in the kerf and not produce a clean cut. On the other hand, thermal cutting processes are geared to the cutting of large format sheets and semi-finished products are only used to a small extent for copper materials in the building of apparatus and ships.

Plasma cutting with a plasma of argon and hydrogen is possible for copper-based materials. The material is melted by the hot plasma jet and the combustion product immediately blown out of the gap while the tungsten electrode is only consumed slightly by evaporation. The process has been used successfully e.g. for cutting large sheet sections of CuNi10Fe1Mn or of steel sheets with CuNi10Fe1Mn cladding. The sheet thicknesses were in the region of 6 to 20 mm, the intersections were smooth and clean and the heat affected zone very narrow.

The thermal cutting of copper materials under shielding gas with a self-consuming steel wire electrode and laser jet cutting is also possible. Both processes are hardly ever used in practice. As the use of these processes for copper material is not recognized in automotive engineering, these processes will not be considered further.

7.2
Joining

Among the joining processes, besides positive substance joints (soldering, brazing, welding, bonding) there are also mechanical joining processes such as screwing, riveting and folding. Screw joints are distinguished only in the strength from those of other metals and riveting and folding are of no significance for copper-based materials in automotive engineering. Hence we will only consider soldering, brazing, welding, plating and bonding in the following. For copper materials, soldering and brazing besides MIG and TIG welding are the most important.

7.2.1
Soldering and Brazing [37]

Caution: Cadmium is banned in at least parts of the automotive industry

Besides its outstanding conductivity values, copper has its excellent solderability and brazeability to thank for becoming the most important material for electrical and electronic applications. Even though soldering is more important here, suitability for brazing is a prerequisite for many applications.

By definition, soldering and brazing are thermal processes for positive material joining and coating of metallic materials whereby a liquid phase occurs by melting a filler metal below the melting temperature of the components to be joined or by diffusion on the interfaces. Base metals are wetted, without themselves melting; flux material is usually added to improve the wetting. A distinction is made between soldering (below 450 °C) and brazing (above 450 °C) according to the liquidus temperature of the filler metal. The working temperature of the filler is the lowest temperature at which the filler wets the base metals. This temperature is normally above the filler's liquidus temperature. Soldering and brazing are therefore different from welding in which the base metals are melted and the melting temperature of the welding filler is roughly the same.

The soldering and brazing processes differ in the heat source. It is necessary to ensure that the working temperature of the filler is reached as quickly as possible. The soldering iron, electric or gas heated, is only used for soldering. It is however not suitable for gap soldering with long overlappings. Flame soldering or brazing can be used for all working temperatures. Blowtorches with city gas/air, acetylene/air or propane/air mixtures are used. Fuel gas/oxygen mixtures are the most economical solution for thick cross sections. Dip soldering is only used as soldering process. For example, wave soldering (wave or drag soldering) is used for soldering printed circuits. Furnaces are used for both soldering and brazing. For example, automotive radiators are soldered in the furnace, newly brazed in a furnace with a rather low-melting silver- and cadmium-free copper-base filler metal. The process is known under its trade mark *Cu*proBraze® and is becoming established for small series production, demonstrating potential in passenger car production, for the aftermarket, for off-road trucks and construction machines and for industrial applications. Applications include heat exchangers, charge air coolers, oil coolers and HVAC. Brazing under protective gas and in a vacuum is possible with such copper materials which are free of evaporable alloy elements. Protective gas or vacuum furnaces are not recommended for brazing copper–zinc alloys or soldering with zinc or cadmium-containing filler metals. Besides these often used processes, electrical resistance and induction soldering and brazing are gaining importance.

7.2.1.1 **Soldering**

The chemical composition and delivery forms of filler metals for soldering are standardized in EN 29453; this contains besides tin–lead and tin–lead–antimony solders, other solders such as tin–antimony, tin–lead–bismuth, tin–indium, tin–silver(–lead) etc. The liquidus temperatures, and therefore also the working temperatures of the solders, are dependent on the weight proportions of both alloy metals. Of the tin–lead and tin–lead–antimony solders, the eutectic alloys S-Sn63Pb37 and S-Sn63Pb37Sb (Sn=62.5 to 63.5%) have the lowest liquidus temperature at 183 °C. Other solders with antimony, bismuth, indium, etc. are used for high or low application temperatures or for second solderings.

Copper materials generally have an oxide layer when ready for shipping. For problem-free soldering, this oxide layer should be completely removed by a suitable cleaning process (see Section 7.4.1). After cleaning, storage times until soldering should be avoided as far as possible.

Two contradictory demands are placed on flux materials. On the one hand, the flux material must remove oxide films still present on the previously cleaned or pickled metallic surface and prevent build-up of new films in order to achieve a high soldering speed with correspondingly low cost but on the other hand, flux materials fulfilling this demand are highly corrosive. They consist essentially of zinc chloride and ammonium chloride and if not thoroughly washed off, the residue at the soldering area and its surroundings can lead to corrosion. As these places are often inaccessible, there is the counter claim for flux materials that they certainly do not pose any corrosion danger. Colophony fulfils this demand but reduces the soldering speed considerably. It can be increased again by adding activation additives to the flux materials. Flux materials according to EN 29454-1 can be divided into three classes according to the corrosiveness of the residues.

- **3.2.2**, **3.1.1** and **3.2.1**, the residues of which cause corrosion as they contain phosphoric acid or ammonium chloride or other acids in an aqueous solution. The residues are to be removed carefully.
- **3.1.1**, **3.1.2**, **2.1.3**, **2.2.1**, **2.2.3**, **2.1.1**, **2.1.3**, **2.2.3**, **2.1.2**, **2.2.2**, **1.1.2**, **1.1.3** and **1.2.2** are flux materials, the residues of which can cause limited corrosion. If and to what extent flux material residue is to be removed must be decided case-by-case.
- **1.1.1**, **1.1.3**, **1.2.3** and **2.2.3** are flux materials, the residue of which is not corrosive. The flux material residue can remain on the workpiece. These non-corrosive flux materials consist of natural, modified natural resins or organic elements and contain no or only halogen-free activating additives.

For applications in electronics and in communications technology, the latter flux materials are of importance as an otherwise necessary washing process can be eliminated. Furthermore, for safety parts, steps are taken to prevent so-called "cold soldering defects".

Unalloyed copper and the aluminum-free copper alloys hardly cause any trouble with soldering. Special flux materials are however normally required for copper–aluminum and aluminum-bearing copper–zinc alloys to achieve problem-free soldering. Oxide films and compounds on the surfaces of these materials are highly resistant but must be removed before soldering. The solderability of the copper materials is tested by various methods. There are also differences in the solderability of aluminum-free copper alloys. The solderability decreases compared with unalloyed copper with increasing contents of tin, zinc and nickel. It must also be borne in mind that tarnish films can form, affecting solderability in the case of some alloys after a long storage period.

The strength of soldered joints is low. The shear strength of capillary gap joints depends on the size of the gap and the strength of the solder which is typically 10 to 40 N/mm^2. However, brazing is the preferred method of joining for high mechanical or thermal loads.

7.2.1.2 **Brazing**

For brazing copper and copper alloys copper–phosphorus, copper–nickel–zinc, copper–zinc or silver-bearing filler metals of silver–copper–zinc and silver–copper–zinc–cadmium with relatively low working temperatures according to EN 1044 are used. The cadmium-containing filler metals have lower working temperatures but should be avoided wherever possible due to the health risks. Filler metals of brass and nickel silver have working temperatures above 800 °C, which can be a disadvantage. This is why silver-bearing and phosphorus-bearing filler metals are preferred as a low working temperature is normally more economical and technically advantageous. Copper–phosphorus filler metals are free flowing and can therefore be brazed without flux materials provided the materials to be brazed have been prepared properly. The melting point of copper is reduced, by the addition of phosphorus, to the eutectic temperature of 710 °C at 8.4% P. The tendency to embrittlement and the working temperature is reduced to 650 °C by adding silver to the copper–phosphorus system while maintaining the free flowing properties.

The advantage of brazing is the higher mechanical strength of the joint that can be achieved but there are also disadvantages. The high working temperatures annihilate the work hardening of cold formed parts. As the strength of all brazings is higher than that of unalloyed soft annealed copper, brazed parts break beside the joint under high load. The hardening of age-hardenable materials can be annihilated by the high heat impact at very high working temperatures. Age-hardenable materials such as copper–chrome–zirconium in an overaged condition will then possess approximately the strength of unalloyed copper in the soft state. Microstructural changes or distortion can appear on the workpieces under some circumstances due to the high brazing temperatures.

Like soldering, the base metal is not actually melted during brazing but, due to the high temperatures, marked diffusion zones are formed between the filler and the base metal that make for a very good joint. Brazing can be carried out under a naked flame, in the furnace, under normal atmosphere, with shielding gas etc., with zinc-free alloys also in a vacuum. Zinc-bearing alloys are not suitable for vacuum soldering due to the zinc's high vapor pressure.

The flux materials are graded according to the working temperature of the filler metals. Of the flux materials listed in EN 1045, the following types are generally suited for copper and copper alloys:

- **FH10** are flux materials with an effective temperature above 550 °C at approx. 800 °C (brazing temperature above 600 °C) containing simple and complex fluorides besides boron compounds.
- **FH11** are flux materials with the same properties as FH10 that also contain chloride.
- **FH21** are flux materials with an effective temperature above 750 °C at approx. 1000 °C (brazing temperature above 800 °C). They are based on boron compounds.
- **FH40** are flux materials with an effective temperature above 600 °C to approx. 1000 °C that do not contain boron compounds.

Pure borax should not be used as flux material as the flux material mixtures are more effective and can be adapted better to the brazing temperatures. They also leave no residue and can be removed more easily.

7.2.2
Welding [14, 59, 60]

Welding is the joining of materials in the welding zone using heat and/or force with or without welding filler materials. In contrast to soldering and brazing, the base metal is melted for welding. The welding process is made possible or facilitated by welding aids, e.g. shielding gas, welding powder or pastes. Two types of welding can be distinguished: joint welding for joining workpieces and deposition welding, i.e. the application of hard-facing, cladding or buffer layers.

Copper and copper alloys can be welded by various processes but not all materials are suited for all processes and not all materials are equally suited. There are generally a number of special points to be noted when welding copper-based materials. Arc welding under shielding gas (MIG or TIG welding), resistance welding (spot, resistance seam and butt welding) and a number of specific welding processes for special applications have become the most important methods of welding copper-based materials. As mainly small workpieces of copper and copper alloys, such as connectors and conductive springs and relays, are used in automotive engineering, the special processes are more important in this field.

The weldability of pure, unalloyed copper is strongly influenced by the material's oxygen content. Oxygen-containing copper types are not weldable or have only limited weldability due to the danger of damage to the weld joint because of brittle phases (Cu–Cu_2O eutectic) and crack formation caused by the reaction of Cu_2O with hydrogen ("hydrogen disease", hydrogen embrittlement). Besides oxygen content, the weldability of pure copper is influenced by high heat conductivity, the tendency to absorb gas (oxygen, hydrogen, carbon dioxide, etc.), major thermal expansion and the transition from liquid to solid and vice versa without a melting interval (solidifying at the melting point). Heat dissipation at 1000 °C is approx. 10 to 15 times that of steel so that it is normally necessary to preheat the workpiece. Largely similar behavior is seen for the highly conductive low alloyed copper materials. Of the pure copper types, Cu-OF (oxygen-free and not deoxidized) is the most suitable for electron beam welding in vacuum.

Copper alloys in comparison with pure copper types are, in principle, better to weld due to the lower heat conductivity. However, the danger of zinc evaporation from copper–zinc alloys, the tough Al_2O_3 oxide layer occurring when welding copper–aluminum alloys and the danger of gas absorption of most copper alloys are to be considered. However, for the alloys there is no danger of "hydrogen disease".

Gas welding, previously used for copper and its alloys, has been replaced almost entirely by arc welding under shielding gas (the MIG and TIG methods) and is used today only in very rare cases.

Achievable strengths of the weld seams are nearly the same as those of the base metals in a soft condition.

7.2.2.1 Arc Welding and Shielding Gas Welding (TIG and MIG Welding)

Open arc welding is not suitable for copper or copper–zinc alloys as the copper's heat conductivity is too high and considerable zinc evaporation occurs with brass. An arc with non-flux-covered metal electrodes will not create a sufficient welding puddle. This is also not achieved by increasing the welding voltage. Open arc welding has a certain significance for the welding of copper–tin, copper–nickel and especially copper–aluminum alloys. Here too it is not possible to avoid pore formation and slag inclusions without preheating the workpieces to between 200 and 300 °C.

The arc welding process under a shielding gas is used for all copper-based materials due to its several advantages. It is mainly the tungsten inert gas (TIG) process in its version as argon arc welding that has become established for practically all copper materials. Zinc evaporation must be observed for materials such as brass, special brass, nickel silver, etc.

In the TIG process, the arc burns freely between the tungsten electrode that is hardly being consumed and the workpiece under a shielding gas of argon or an argon–helium mixture. Single layer welding is used for wall thickness up to 4 mm. It is only stated for the sake of completeness that the metal inert gas (MIG) process, in which the arc burns between a consumable electrode (welding filler, wire electrode) and the workpiece, is particularly suited for greater material thicknesses. The MIG process is especially recommended for copper, copper–tin, copper–nickel and copper–aluminum alloys. The evolution of zinc fume results in welds which are likely to be porous and consequently unacceptable, particularly if autogenous welding is attempted. Zinc fume also makes visual observation of the welding operation difficult. Caution is therefore advised when welding brasses.

In the classification of the copper welding filler materials according to EN, there is no connection between the product form of the welding filler and the welding process.

Welding fillers (solid wires and rods) for fusion welding of copper and copper alloys are standardized in EN 14640. The compositions, requirements on the properties and limiting dimensions for rods and wires of copper and copper alloys for welding and brazing filler materials are laid down in EN 13347.

Unsuited in any case are welding fillers of the base metal, e.g. strips of copper, as they produce porous weld seams. As pure copper has a melting point and no melting interval it is difficult to shape the cross section of the weld and exclusively alloyed fillers are used also for unalloyed copper. Notes on the selection of suitable filler materials for various base metals are contained in the relevant specifications and can also be found in the DKI brochures [59, 60].

Flux materials based on boron compounds with the addition of easily oxidised dissolving metal salts are suited for copper and most copper alloys. Special fluoride-containing flux materials are required for copper–aluminum alloys to dissolve the refractory oxide skin. The wearing of a respirator and efficient exhaust of the vapors of such flux materials are necessary.

7.2.2.2 **Resistance Welding**

In resistance welding, the necessary heat is produced by an electric current via the electrical resistance of the welding area. This means, that the otherwise highly undesirable *copper losses* $= I^2 \cdot R$ of electric machines are used to produce the heat for welding. Welding is carried out here with or without force impact and with or without filler material. These processes are metallurgically significantly simpler than pyrometallurgical processes. On the other hand, the workpiece's high electrical conductivity can limit the weldable cross sections.

Due to the normally high conductivity of the copper materials, resistance welding processes can only be carried out on high performance machines enabling accurate control of contact pressure, welding current and welding times. Spot, resistance seam and butt welding processes have become more important for copper and copper alloys, especially for automated mass production of small parts.

Resistance welding of unalloyed copper for electrical applications has become more important as the low oxygen content does not affect weldability. Spot and resistance seam welding of unalloyed copper is limited to overlap joints compressed by stamp- or wheel-shaped electrodes, whereby the current impact raises the temperature to above the melting point.

Copper alloys can generally be easily resistance welded due to the lower conductivity if the surfaces of the pieces to be joined are chemically clean. Resistance welding processes are mainly used for thin strip copper materials for mechanized or automated production.

The flash welding process is used for welding copper cords and copper ropes.

Special resistance welding processes are the high-frequency resistance pressure welding and the impulse current welding process (the ultrapulse or short-term welding process). In the first process, the current displacement effect of high frequency current (300 to 500 kHz) is used, by which the welding heat is concentrated onto the faces. This is particularly advantageous for high conductivity materials such as unalloyed copper. This process is excellently suited for seam welding thin walled tubes and profiles. Seam welded copper tubes of 12.5 mm ∅ and a wall thickness of 0.127 mm can be welded at a speed of 92 m/min. Development was carried out to use such tubes for the production of lightweight, high-strength, high efficiency automobile radiators (*Cu*proBraze®). The second process is used for difficult to weld small and jewellery type parts in microtechnology and electronics for which exact heat management is necessary. Welding heat results from a power pulse with a peak of 1500 to 3500 A from a capacitor discharge with a pulse duration of 3 to 6 ms. This process enables the joining of copper with other metals of higher conductivity.

As electrode materials for spot welding steel and non-ferrous metals, copper, copper–chrome–zirconium, copper–cobalt–beryllium and as a new development, copper–nickel–silicon–chrome can all be considered, depending on welding performance. Data sheet DVS 2903 "Elektroden für das Widerstandsschweissen" ["Electrodes for resistance welding"] from the German Association of Welding Technology DVS [*D*eutscher *V*erband für *S*chweißtechnik] holds some information on this subject.

7.2.2.3 **Recent Welding Processes**

Special welding processes are used for special welding problems. Ultrasound welding for applying precious metal contacts onto copper alloy relay springs, TIG and MIG pulsed arc welding for deposition welding of copper onto steel (weld cladding), electron beam welding in a vacuum and laser beam welding in atmosphere under shielding gas.

Spot and resistance seam joints on overlaps of thin sheet and strip for practically all material combinations are possible with ultrasound welding. With TIG or MIG pulsed arc welding, the degree of dilution, i.e. the penetration of copper into the steel matrix (solder brittleness!), is significantly reduced. Cu-OF is easily weldable by electron beam welding while other copper types are less weldable and copper alloys with the exception of zinc-bearing alloys (brass, special brass and nickel silver) are very suitable. Using this process, copper materials can be joined to carbon steel and low alloyed steel, to corrosion resistant steel, to nickel and to nickel and cobalt alloys. With laser welding of copper and copper alloys, the significantly higher reflectivity in comparison with steel must be considered.

The reduced absorption of the beam energy reduces the possible welding depth. Copper alloys, with the exception of brass, are better suited than copper for this process. This last process is used for welding copper–beryllium spring wires with resistance wires, of spring wires of copper alloys with gold contacts and lacquered copper wire with steel.

The DKI brochures [59, 60] contain information for the special processes mentioned above and for the following more recent welding processes for copper and copper alloys:

Cold-pressure welding, cold extrusion pressure welding, friction welding, diffusion welding, tungsten plasma welding and microplasma welding.

7.2.3 Plating

The widest variety of metal combinations can be produced by plating with copper-based materials. For metallurgical reasons, this may require intermediate plating, e.g. nickel coating as a diffusion barrier. According to the process, we can distinguish between roll-bonding, explosive plating and deposition welding. All processes are used both for copper and copper alloys.

For roll-plating, the base metal and cladding materials, e.g., steel as basic material and copper–nickel as cladding, while maintaining certain temperatures and pressures, are rolled on heavy thick metal plate rolling mills. The heat treatment necessary for the material combination follows after rolling. Then, the plating layer is ground to the surface finish required for the intended application.

For explosive plating, both the base metals and the cladding materials, normally of similar dimensions, are arranged on the blast bed on top of each other at a defined distance with their bare metallic surfaces facing one another. The layer of explosive is applied to the surface of the cladding material and fired in a punctiform or linear manner. The cladding material is accelerated towards the base metal and crashes at great speed onto it. Both materials are fused to one another by the high pressure. This process also enables the joining of materials which cannot be joined by rolling or welding, e.g. titanium layers onto copper as base metal.

Most frequently, steel of the widest variety of compositions has been used as base metal. Typical copper-based cladding materials are Cu-DHP, CuNi10-Fe1Mn, CuNi30Mn1Fe, CuAl8Fe3 and CuZn39Sn1, but copper–aluminum composites, e.g. for contact rails, have also been produced.

Cladded parts which need a high degree of deformation for their production sometimes cannot be produced by explosive or roll plated sheet or strip because they are too hard after plating. In this case deposition welding is the solution for cladding these parts after forming. The MIG or TIG pulsed arc process is preferred for these welds.

7.2.4
Bonding [61]

Copper and copper alloys are well suited for bonding. In order to produce high quality adhesive joints, surface preparation is most important (see Section 7.4.1). Copper material surfaces are best suited for bonding after machining. The application of an intermediate layer, e.g. a chromate layer, as a primer is recommended under special circumstances. If painted surfaces are to be bonded, the bond strength of the paint should be checked beforehand.

Great adhesive strengths can be achieved with adhesive bonds of copper or copper–zinc alloys with aluminum, stainless steel and carbon steel with dual component polyurethane-based adhesive. The DKI brochure [61] contains details on the adhesives suitable for copper materials, bonded joints and achievable strengths etc. As copper materials are easily soldered, bonding technology for copper materials was only proven at a relatively late stage. The advantage of bonded joints is mainly established by the fact that not only can metals be joined with each other but also with other materials, e.g. coatings, wood, synthetics, etc. On the other hand, and in contrast to bonding, soldering needs no setting time. The strengths of bonded joints are of the order of those of soldering.

There are some applications for bonded joints in automotive engineering. For example, the die cast aluminum covers on fuel filters, the shells of which are copper-plated free-machining steel, are bonded. Another example is fuel pumps with copper fuel pipes fixed in place. In the production of tyres, the embedded steel wires used for strengthening are given a coating of brass as an adhesive agent for bonding the steel wires with the rubber.

7.3
Heat Treatment

Homogenization, soft annealing (recrystallization annealing), stress-relieving and age-hardening consisting of solution annealing, quenching and tempering (aging) are customary heat treatments for copper and copper alloys. Due to the variety of copper-based materials and the dependence of temperatures and times of individual heat treatments on the material's composition, microstructure or strength after cold working etc., it is only possible here to go generally into the individual heat treatments.

Reference values for heat treatment temperatures are included in various handbooks [14, 24, 29], data sheets [32, 43] and publications [33, 50]. Heat treatment takes place preferably in electric furnaces with a temperature control accuracy of $\pm 2.5\,^{\circ}$C but can also be carried out in gas furnaces. Annealing is carried out depending on the material and demands under a normal atmosphere or in a vacuum whereby the protective gas is adjusted depending on the demands, i.e. neutral, oxidizing or reducing. Heat treatment can either be carried out discontinuously in batches in a pot, hood type or chamber furnace or continuously in a – mainly horizontally arranged – continuous furnace. In the case of spring strips, on which the strictest demands are placed on uniformity of the microstructure, accuracy (tolerances) and flatness of the strip, heat treatment is carried out in a magnetic levitation furnace (type: Junkers) almost exclusively under protective gas.

7.3.1
Homogenization

Homogenizing takes place exclusively at the semi-finished goods producers or foundries prior to shipping the castings or hot worked semi-finished products to the fabricator. Heat treatment takes place to equalize segregation or local differences in the composition. This essentially concerns alloys with the elements tin (tin bronze with a broad solidification range) and copper–nickel alloys. Homogenizing takes place at relatively high temperatures with long times and is therefore very expensive.

The wrought alloy CuSn8 particularly tends toward strong segregation and is preferably homogenized at 760 °C prior to cold rolling to dissolve the brittle and high tin-containing phase. Homogenizing is not necessary for other copper materials that are hot and cold rolled or intermediately (soft) annealed between individual cold formings.

For age-hardenable copper alloys, solution annealing above the alloying element's solubility limit is also described as homogenizing annealing.

7.3.2
Soft Annealing (Recrystallization)

A metal must be soft annealed if by cold working it assumes such high strength, with a simultaneous reduction in elongation, that cold working cannot be continued. Such annealing generally aims at achieving a soft state (recrystallization) characterized by the formation of new crystals.

Recrystallization depends on the grain size of the microstructure, the deformation degree of the previous cold working and the recrystallization temperature. With too little cold working, there is the danger of formation of coarse grain by annealing. The combinations are shown in a three-dimensional diagram and explain why for a certain material, not one recrystallization temperature but only a temperature range can be given. The temperatures depend on the annealing time. A short annealing time is normally applied for a high temperature; a long annealing time at a low temperature. If there is a recrystallization diagram the temperature for a certain annealing time can be given if the degree of deformation and the initial grain size are known. Material regeneration precedes recrystallization. This means the regeneration of a material without new grain formation. Grain growth follows after recrystallization if the temperature is maintained for too long. It is a growth of energetically favorable

Table 36 Recommended times and temperatures for soft annealing of a number of selected copper-based materials.

Symbol	Temperature (°C)	Time (h)	Symbol	Temperature (°C)	Time (h)
Cu-ETP/Cu-FRHC	300–650 [a]	0.5–3	CuSn8 [c]	500–700	0.5–3
Cu-OF	425–650	0.5–3	CuNi18Zn20 [d]	600–750	0.5–3
Cu-PHC/Cu-HCP	350–650	0.5–3	CuNi18Zn27 [d]	600–750	0.5–3
Cu-DHP	350–650	0.5–3	CuNi9Sn2	600–700	0.5–3
CuZn15	425–650	0.5–3	CuNi10Fe1Mn	625–750	0.5–3
CuZn30	450–675	0.5–3	CuAg0.10	400–650	0.5–3
CuZn33	425–700	0.5–3	CuAg0.10P	400–650	0.5–3
CuZn37	450–600 [b]	0.5–3	CuFe2P	650–700	0.5–3
CuZn36Pb3	425–600	0.5–3	CuSP	425–650	0.5–3
CuZn39Pb2	425–600	0.5–3	CuTeP	425–650	0.5–3
CuZn39Pb3	425–650	0.5–3	CuZn0.5	425–600	0.5–3
CuZn40Pb2	425–650	0.5–3	CuBe2	720–760	0.5–3
CuZn31Si1	500–600	0.5–3	CuBe2Pb	720–760	0.5–3
CuZn38Mn1Al	500–650	0.5–3	CuCo2Be	920–960	0.5–3
CuZn37Mn3Al2PbSi	500–650	0.5–3	CuNi2Si	725–760	0.5–3
CuSn4 [c]	500–700	0.5–3	CuCr1Zr	600–800	0.5–3
CuSn6 [c]	500–700	0.5–3	CuZr	850–965	0.5–3

a) When annealing in reducing atmosphere, the temperature must be kept below 450 °C, to prevent hydrogen disease (embrittlement).
b) It is important not to exceed the maximum temperature so as not to deteriorate the material's cold forming properties.
c) The alloys are prone to cracks when annealing and should be thermally stress-relieved beforehand.
d) The alloys must be annealed in an oxidizing atmosphere. They are prone to cracks and should be thermally stress-relieved beforehand.

grains at the expense of less favorable grains. Table 36 shows soft annealing temperatures for a number of copper materials as reference values.

7.3.3 Thermal Stress-relieving

Stress-relieving reduces residual stress in a semi-finished product or workpiece caused by cold working without leading to too great a drop in strength. Stress-relieving is most important for copper materials tending towards stress corrosion cracking, e.g. for copper–zinc alloys. For stress corrosion cracking, beside a corrosive medium and a sensitive material, the existence of residual or load stresses is necessary. So the risk of stress corrosion cracking can be considerably reduced by relieving the residual stresses from production.

Table 37 Recommended annealing times and temperatures for stress-relieving of selected wrought copper-based materials.

Symbol	Temperature (°C)	Time (h)	Symbol	Temperature (°C)	Time (h)
Cu-ETP/Cu-FRHC	150–200 [a)]	1	CuSn8	200–300	1
Cu-OF	150–200	1	CuNi18Zn20	300–400	1
Cu-PHC/Cu-HCP	150–200	1	CuNi18Zn27	300–400	1
Cu-DHP	150–200	1	CuNi9Sn2	250–400	1
CuZn15	200–300	1	CuNi10Fe1Mn	280–450	1
CuZn30	200–300	1	CuAg0,10	250–300 [a)]	1
CuZn33	200–300	1	CuAg0,10P	250–300	1
CuZn37	200–300	1	CuFe2P	200–300	1
CuZn36Pb3	200–300	1	CuSP	150–200	1
CuZn39Pb2	200–300	1	CuTeP	150–200	1
CuZn39Pb3	200–300	1	CuZn0,5	200–300	1
CuZn40Pb2	200–300	1	CuBe2	250–300	1
CuZn31Si1	250–350	1	CuBe2Pb	250–300	1
CuZn38Mn1Al	300–430	1	CuCo2Be	350–420	1
CuZn37Mn3Al2PbSi	350–450	1	CuNi2Si	350–450	1
CuSn4	200–300	1	CuCrZr	300–350	1
CuSn6	200–300	1	CuZr	350–400	1

a) When annealing in reducing atmosphere, the temperature must be kept below 450 °C, to prevent hydrogen disease (embrittlement).

Thermal stress-relieving also increases dimensional stability by reducing the danger of warping caused by residual stress. Additionally residual stress adds to external load so that the fatigue limit of a workpiece may deteriorate.

For thermal stress-relieving, the temperature and annealing time are selected so that the recrystallization threshold is not reached. Material regeneration takes place, lattice defects are cured and stress relieved without significantly reducing strength properties. Thermal stress-relieving should take place following cold working, which may have created residual stress in the workpiece. For example residual stress can be caused in the workpiece by machining. Table 37 shows recommended temperatures and annealing times for thermal stress-relieving of a number of selected copper materials.

Standardized copper casting materials, except copper–aluminum castings, are stress-relieved at 250 °C. Annealing time is based on 1 h per 25 mm wall thickness. The annealing temperature of copper–aluminum casting alloys is 315 °C.

7.3.4
Age-hardening (Precipitation Hardening)

A number of copper alloys are age-hardenable. Three conditions must be fulfilled for the hardening of copper alloys (and all metal alloys):

- There must be a limited solid solubility for the alloying element.
- The solubility must decline with decreasing temperature.
- The speed of reaching the equilibrium must be so low that the homogeneous solid solution of the higher temperature can be frozen by quenching.

Figure 15 explains the hardenability of copper alloys.

If annealing in temperature interval 3 is carried out long enough, the atoms of the alloy partner will be dissolved in the copper lattice until a uniform α-mixed crystal remains (*solution annealing, homogenizing*). This state must be "frozen" by quenching the material in order to be able to harden it. The alloy is in a soft condition after quenching (very unlike hardenable steel).

The material is then age-hardened by tempering. The workpiece is heated to temperatures at which diffusion of the alloy atoms and thus segregation in accordance with equilibrium solubility is possible (*age-hardening*). The temperature must be set low enough that the particle sizes of the precipitates remain in the sub-microscopic range (see region 5 of the diagram). Hardness and strength are higher, the more finely the segregations are distributed. This mechanism leads to increased hardness and to increased electrical conductivity.

At higher temperatures (e.g. region 4 of the diagram) hardness, strength and conductivity are again reduced as the precipitates coagulate and the segregated quantity is reduced again due to increasing solubility. Too long annealing times

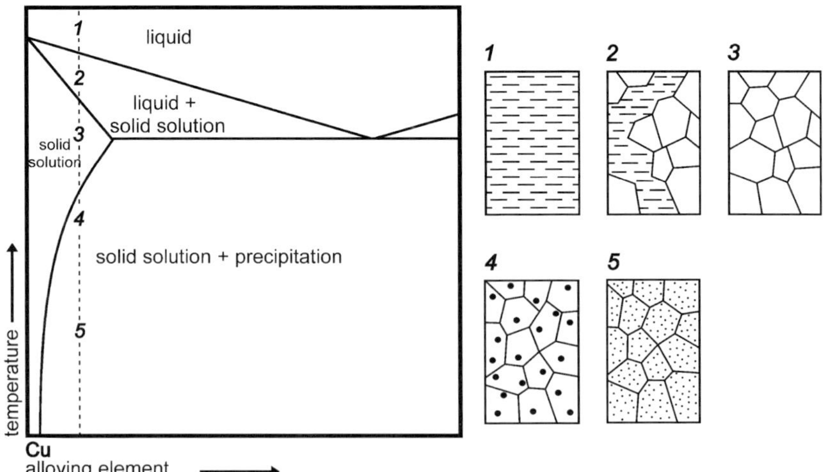

Fig. 15 Phase-diagram showing the condition of the age-hardenability of (copper) alloys.

have a very similar effect (over-aging). If the precipitates can be seen in an optical micrograph, over-aging has already taken place.

The mechanism of precipitation hardening offers some advantages for the user. In the soft (solution treated) state, as delivered by the producing mill, the workpieces can be easily processed and cold formed (in some cases it may be advantageous to age-harden the workpieces to get shorter chippings when machining). The following age-hardening process is to be handled easily by the manufacturer of the finished workpiece, as warping is not to be expected to a large extent because of the low age-hardening temperatures (age-hardening is at the same time a kind of stress-relieving).

The strength-increasing effect of age-hardening is enhanced if there is cold-working between solution treatment and age-hardening.

Table 38 Heat treatment for a number of age-hardenable copper alloys.

Materials and conditions		Solution annealing – Homogenizing		Hardening	
		Temperature (°C)	Time (min)	Temperature (°C)	Time (h)
CuZr	solution annealed and tempered	900–930	5.5	500–550	1–4
	solution annealed, cold worked and tempered			375–470	
CuCr1Zr	solution annealed, work hardened and tempered	900–1000	1–2	400–500	1–2
CuBe2	solution annealed and cold aged	770–800	10.5	315	3
	solution annealed and work hardened, ¼ hard				2
	solution annealed and work hardened, ½ hard				
	solution annealed and work hardened, hard				
CuCo2Be	solution annealed and cold aged	910–940		480	3
	solution annealed and work hardened, ½ hard				2
	solution annealed and work hardened, hard				
CuCr1	solution annealed and tempered	980–1010		425–550	2–4
	solution annealed, cold worked and tempered				
CuNi2Si	solution annealed and drawn	745–800		455–480	1.5
CuCr1-C[a)]		1000–1010	1	480	3

a) For castings, times for solution annealing and for tempering are in hours (h) per 25 mm wall thicknesses.

The standards contain the following age-hardenable alloy systems: copper–chrome, copper–zirconium, copper–chrome–zirconium, copper–beryllium, copper–cobalt–beryllium, copper–nickel–beryllium and copper–nickel–silicon.

Table 38 shows the temperatures and annealing times for age-hardening of a number of age-hardenable copper alloys.

7.4 Surface Treatment

Surface treatment on copper materials is carried out to clean the surfaces, to improve their appearance or to prepare them for a subsequent process, for instance a joining process or the application of a decorative coating or a protective coating against wear or corrosion. Please note that the best coating is worthless if it does not stick fast to the surface.

7.4.1 Pretreatment [14, 62]

Various pretreatment processes are used for copper materials. Mechanical, chemical and electrolytic surface treatment processes are described in the DKI brochure [62] and the literature [14] and these processes can only be dealt with briefly here. The selection of processes depends on the surface requirements, often caused by the following process.

7.4.1.1 Cleaning, Degreasing

The simplest cleaning processes are mechanical methods such as brushing, scraping or blasting. For the first two processes, rotating brushes, in the first case with copper or copper alloy wires of various hardness, for scraping, the finest corrugated wires of the same materials are used as residue of the wires could otherwise cause contact corrosion. Only in the case of severe contamination, e.g. for removing foundry sand, are steel wire brushes used beforehand. Blasting is used to remove sand residue from castings, scale or oxide layers with natural or synthetic substances, with steel shot, cut steel wire, grit of non-ferrous metals and slag. It is to be borne in mind that the relatively soft surface of copper materials may be disturbed.

For degreasing surfaces of copper materials, either organic solvents or alkaline cleaning agents or combinations of both are used. Alkaline solutions are preferred if copper-based materials are to be electrolytically degreased. Combined degreasing processes are required for especially severe contamination.

Organic solvents dissolve oil and grease but not persistent industrial dirt and other inorganic elements. Chlorinated hydrocarbons such as trichloroethylene, perchloroethylene, trichloroethane and mixed solvents are still in use although their production is gradually being discontinued due to environmental aspects.

The chemical industry meanwhile offers substitutes. Degreasing can take place by immersion, but degreasing in the vapor phase whereby the solvent condenses on the surface is much more effective. These processes must however be carried out in closed installations which ensure zero escape of the solvent into the atmosphere and allow recovery of the solvent. The use of ultrasound intensifies the cleaning effect.

However, degreasing in a hot, aqueous alkaline solution, often described as degreasing by boiling, is the most effective.

The degreasing process can be accelerated and controlled by the use of direct current. Electrolytic degreasing is used as the preferred method of cleaning prior to electrolytic coating. The same solutions are used as for chemical degreasing. Recipes are contained in the literature [14, 62].

Whether or not the grease has been removed can easily and quickly be checked by a wetting test. It responds to the slightest trace of grease.

7.4.1.2 **Pickling and Etching**

Pickling in acids or acid mixtures is used to remove the scale, oxide layers, soldering slag and tarnish films. The objective of pickling is to produce a chemically clean and/or matt or shiny surface. Surfaces must be cleaned and completely free of dirt, oil and grease prior to pickling. Previously, only sulfuric acid was used for pickling. Other, newly developed pickling solutions have nowadays become more common. These new pickling solutions were developed in order to control wastewater more efficiently and, in particular, to enable recovery of the metal from the pickling solutions. Modern pickling installations are operated in closed loops for ecological reasons.

The DKI brochure contains recipes and recommendations for pickling baths consisting of sulfuric acid, ammonium persulfate, amidosulfuric acid and fluoroboric acid.

Copper alloys can be pickled under ultrasound for which sulfuric acid is normally used.

Burnishing involves pickling to obtain a matt or shiny surface. Burnishing has become less important due to the development of newer pickling agents, as these pickles can achieve similar results to burnishing. The DKI brochure contains detailed information.

Etching involves removing part of the metal, e.g., to make the material's grain structure visible, for bringing decorations to the surface or in the case of printed circuits, to remove material except the conductor paths. Copper can be chemically and electrolytically shallow or deep etched using various solutions. Etching agents can also work in a passivating manner. Potassium and ammonium persulfate solutions are used for light etching and dichromatic solutions for deep etching, e.g. for the production of printed circuit boards. Further information is included in the literature mentioned above.

7.4.1.3 Chemical and Electrochemical Polishing

Chemical or electrolytic polishing or shining achieves better surfaces than pickling.

Chemical polishing as opposed to electrolytic shining has the advantage of simplicity of the process. Polishing takes place by immersion of the workpiece into the chemical solution and removal after a few minutes. The shine produced in this way is not as good as that produced by electrolytic polishing. The costs of the chemicals are slightly higher than for pickling. The process is described in the DKI brochure that also contains recipes for the chemicals.

For electrolytic polishing, the workpiece is suspended as an anode in a special electrolyte. On electrolysis, microscopically small elevations are preferentially removed, thereby achieving a flattening or polishing effect. The process is also suitable for removing burr. The DKI brochure contains notes on the process. Indeed, not all copper materials are equally suited for this process. For copper–zinc alloys, the polishing effect diminishes with increasing zinc content, i.e., with increasing amounts of β-mixed crystals. It should be kept in mind that the suitability for machining and electropolishing go in opposite directions. A compromise in the selection of material has to be found if both processes are used on a workpiece.

7.4.1.4 Pretreatment for Soldering, Brazing or Welding

The surfaces to be joined must be chemically clean if good joints have to be achieved. Oxide scales have to be removed prior to welding. To clean semi-finished products, it is generally enough to clean the prepared weld joints with wire brushes (with brass or bronze wire). If however, the workpieces are contaminated or contain oil or grease, the previously described cleaning processes must be applied. Contaminated or particularly oily or greasy weld joints can lead to gas absorption in the weld puddle and therefore to porous weld seams. With copper–aluminum alloys, the highly resistant oxide layer must either be penetrated by high frequency impact for igniting an arc or special flux materials must be used for welding. Pickling such material is not sufficient.

The same provisions also apply in principle for soldering and brazing. To complicate matters, the joints on the workpiece are not melted, only wetted. Efficient wetting and subsequent diffusion between material and filler are only possible if the material at the joint is chemically clean. Normally, weak non-corrosive flux material suffices to enable efficient solderability on surfaces cleaned by brushing or pickling. On many copper alloys, tarnish films nevertheless form in the case of longer storage times, deteriorating solderability. Attempts are made to counteract this by using anti-tarnish paper, normally containing benzotriazol-based inhibitors. If solderability is affected by longer storage periods, tarnish films must be removed by pickling or more aggressive flux material which must be completely removed after soldering. Otherwise there is the danger of corrosion on the workpiece.

7.4.1.5 **Pretreatment for Bonding** [61]

The wetting and absorption capacity of the wash primer and thus the adhesive strength of bonded joints can be increased multi-fold by suitable pretreatment of the surfaces to be joined. Pretreatment has the objective of cleaning the surface and increasing the effective bonding area by fine roughening of the surface, thereby improving the activity for bonding.

The bonding surfaces are cleaned with organic cleaning agents such as acetone or butanone or better still alkaline cleaners. Ultrasound improves the effect of solvents. Perchloroethylene is currently still used for vapor degreasing.

While some adhesives, e.g. pressure sensitive adhesives, such as cyanoacrylate and anaerobic adhesives can only be applied on cleaned and completely degreased surfaces, acrylate adhesives and epoxy resins with special additives will even stick fast to oily surfaces.

The easiest way of removing oxide layers and creating a defined roughness is by mechanical means, by machining, by shot blasting, brushing, grinding, sanding or polishing. Blasting has thereby become more established than grinding or sanding by hand. The surface roughness should however not exceed 30 μm as air locked in pores may otherwise lead to imperfections.

7.4.1.6 **Pretreatment for Coating**

Prior to coating and before further chemical treatment, paint residue, layers of oil or grease, the residue of substances from production, foundry sand, scale, tarnish and/or oxide layers must be carefully removed from the surface according to the methods described above. The workpiece must then be pickled, polished and coated. The cleaning and pickling treatment is followed by rinsing, to be carried out with great care. Residues of pickling acid may cause corrosion. All processes, including rinsing, are described in more detail in the DKI brochure [62].

7.4.1.7 **Shot Peening**

The strengthening of surface zones by shot peening with the objective of increasing the fatigue strength of workpieces, as used for other metals, is also used for copper and copper alloys. Glass beads of pure material must be used for peening to prevent foreign matter being embedded into the surface. Each of the microbeads works like a tiny hammer. The surface layer is strengthened to a depth of 25 to a few hundred μm depending on the speed and the size of the glass beads. Additionally shot peening induces compressive stress in the surface zones which may superimpose favorably to the tension stress of exterior load and thus add to the fatigue life of the workpiece (see Section 6.2.6).

The surface, smoothed and strengthened by shot peening counteracts crack formation as nearly all fatigue cracks emanate from the surface and their occurrence is promoted by irregularities such as notches. Parallel to strengthening, flattening of rough surfaces also takes place e.g. grinding scratches can completely vanish.

7.4.2
Surface Design

Copper and copper alloy surfaces can be decoratively styled with a wide variety of processes. For copper and copper alloys the processes for treating surfaces for decorative purposes described in Section 7.4 are more important than for improvement of the wear or corrosion resistance.

7.4.2.1 Mechanical Processes for Surface Treatment

In order to achieve certain surface effects, copper and copper alloy materials can be ground and polished according to the processes described in the brochure [62]. Cold formed materials of fine grain sheets or strips require no pre-grinding. Mechanical polishing is carried out with cloth or felt discs and polishing pastes. Mass production components are ground and polished in a rotating barrel.

Matt finishing and dash-matt surfaces can also be achieved with fiber and nylon brushes. Matting requires wire brushes, the wires of which are harder than the work metal. For graining, i.e. producing a very fine matt finish, brushes with wire clusters are used.

7.4.2.2 Non-metallic Coatings, Metal Coloring [63–66]

Inorganic coatings are formed when coloring copper and copper alloys as well as in enamelling. While enamel coatings, however rarely, are applied for reasons of corrosion protection, coloring is applied purely for decorative reasons. Even though copper materials have their own decorative colour, there is often a wish for other colors.

The very thin layer of dye on copper materials originates from reactions of certain chemicals, normally in aqueous solutions, with the metal surface. The DKI literature contains recipes for chemical coloring and enamelling [63–65].

Organic coatings can be produced with all the usual paint and synthetic resin coatings. Transparent paints are quite normal for copper materials, e.g. in order to preserve the natural hue of the materials or of the chemical colouring. The coating of copper and brass with a wide variety of transparent paints is explained in detail in a DKI brochure [66].

7.4.3
Metallic Coatings

Copper and copper alloys are excellent base materials for electrolytic surface coatings but electroless platings are also common. PVD, CVD and laser coatings are also possible but have not yet been used to a large extent in a technical context as far as the author is aware.

If there is danger that the coating may be damaged and a corrosive medium may appear, we warn against the plating of more noble metals than the base metal (silver, gold on copper; copper on steel, aluminum, magnesium) because of the danger of galvanic corrosion which will dissolve the (possibly load bearing) base metal.

7.4.3.1 **Electroplating**

Electroplating is carried out with the help of electric current, whereby the workpiece is suspended as the cathode in an aqueous solution of metal salts. Nickel and nickel–chromium layers are very popular. The coating thicknesses and combinations are thereby selected for the intended use. These coatings are for decorative purposes and corrosion protection. They are often additionally bright chromium-plated to improve corrosion protection. The hard chrome plating (thickness 20–250 μm, hardness from 850 to 1100 DPH) increases wear resistance.

Besides hot-dip tinplating, electro-tinning on copper and copper alloys is common. In particular, strips are coated with layers of pure tin or tin–lead (e.g. S-Sn60Pb40 according to EN 29453) to improve their soldering properties. These strips are used extensively for electrical and electronic applications. Because of the lower coating speed of electrodeposition this process is mainly used for thin layers of pure tin. The danger of whisker formation in the case of pure electro-tin coatings is a disadvantage compared to hot-dip tinning.

Besides the metallic coatings mentioned above, electrocoatings of silver, gold and other precious metals have gained importance for components for electronics and communications technology. Electrodeposited lead coatings offer protection against sulfurous fuel gases.

Also leaded copper–zinc alloys (free-machining brass) can be electrocoated without any problems if careful pretreatment by pickling in hydrochloric or hydrofluoroboric acid is carried out.

Further information is contained in the worksheet "Einfluss des Grundwerkstoffes Kupfer und seiner Legierungen auf das Ergebnis galvanotechnischer Fertigung" [The influence of the base material copper and its alloys on the result of electroplating] (Deutsche Gesellschaft für Galvanotechnik e.V., Düsseldorf).

For the sake of completeness, we mention that copper and copper alloys can of course be electrodeposited onto other metals. Of technical importance are electrodeposited copper coatings such as running-in liner, protective layers against galling and for repair purposes. Electrodeposited copper layers are widely used as gas-proof coatings and barrier layers. Electrodeposited layers of copper–zinc alloys are used for decorative applications, serve as corrosion protection in interior applications and as sliding layers in bearings and for deep drawing. Layers of brass are also used as the tie layer for rubber coating. Electrodeposited bronze layers are also used for corrosion protection and as sliding layers [67].

7.4.3.2 **Electroless (Chemical) Deposition**

Electroless chemical deposition is also possible on copper and copper alloys. Electroless deposited nickel coatings are used for corrosion- and wear-protection. The hardness of the coating of 550 DPH can be increased to 1000 DPH by heat treatment at 400 °C/1 h. Usual applications are interior coatings of copper tubes with nickel as corrosion protection.

Electroless copper coatings are used for sliding purposes or as repair coatings, mainly on interior parts. They are however not suited as gas-proof coatings or barrier layers [67]. It must be borne in mind that the electroless processes are generally very expensive.

7.4.3.3 **Hot-dip Coating**

These coatings are produced by immersion of the pre-treated materials or the semi-finished products in molten metal baths. Intermetallic phases form at the interface and on extraction the adhering metal solidifies on the surface as a coating. The thickness of the coating is checked by stripping off the liquid film on the surface.

The process is of special importance for copper and copper alloys for hot-dip tinning of workpieces. Strips are coated with S-Sn60Pb40 to improve the solderability. The achievable coating thicknesses are between approx. 1 and 15 μm. With mechanical stripping equipment layers are set between approx. 1 and 8 μm and with air nozzles between 2 and 15 μm.

Besides tin or tin–lead coatings lead or lead alloy coatings are deposited on copper or copper alloys by hot-dipping. Lead coatings offer protection against sulfurous fuel gases [67].

7.4.3.4 **Thermal Spraying**

Until now, thermal spray coatings are of no technical importance for copper or copper alloys [67]. However, copper–zinc alloys (brass) and copper–tin alloys (tin bronze) are sprayed onto other, normally iron, materials as a sliding layer.

7.4.3.5 **Deposition Welding** [68]

Copper and copper alloys are of major importance as deposited materials. In this way, not only on materials of the same type but also on different types of material, e.g. iron and steel parts, defects are repaired and material replaced on worn parts. With copper alloys steel surfaces are protected against corrosion (Caution: If there are defects or damage in the copper coating and the iron surface is laid bare locally there will be accelerated corrosion on the iron), armored against certain kinds of wear and sliding surfaces deposited in many forms. Deposition weldings are also carried out with copper–aluminum alloys as combined protection against corrosion and wear.

Wire and rods for fusion welding of copper and copper alloys are standardized in EN 14640 and rods and wires of copper and copper alloys for welding and braze welding are standardized in EN 13347. For repair purposes and protection against corrosion, the following welding fillers are used according to EN: CuAg1, CuSn1, CuSi2Mn1, CuSi3Mn1 and CuNi30. When mechanical wear as well as corrosion is to be fought, welding fillers of copper–zinc, copper–tin and copper–aluminum alloys are used. When depositing sliding layers, materials such as CuSn6, CuSn12 or CuZn40Si are selected, if the sliding partner is to be spared. If however, the layer is to be the more resistant, harder coatings, e.g. of the heterogeneous Cu-Al-Ni- or Cu-Al-Fe alloys, are applied whereby the hardness can be increased to 250 HB.

When deposition welding, the arc welding process, preferably the MIG and TIG processes under shielding gas are used. Possible alloy formation (commingling of the metals with possible formation of brittle phases) is to be borne in mind when joining different kinds of material by welding.

7.4.3.6 Roll-cladding

Copper materials with roll-clad precious metal coatings are used as materials for contacts. Copper–zinc alloys with high copper content, tin bronze, copper and special copper alloys are used as base materials.

Contour strips for the manufacture of connectors of copper or copper alloys are also produced by roll-cladding of strips (besides longitudinal welding or contour milling or peeling).

7.5 Safety Measures

No special safety measures apply to copper materials such as for instance those for grinding aluminum or other metals. There is no danger of dust explosions with copper base materials. Maximum allowable concentrations (MAC-Value) for smoke are $\leq 0.1\ mg/m^3$ (alveolar fraction) and for grinding dust $\leq 1\ mg/m^3$ of the respirable fraction. When brazing or welding copper alloys evaporation of zinc is to be minded and also nickel and beryllium and vapors of fluorine-containing fluxes. It is possible to keep MAC-limits safely by fume extraction.

7.6 Health Care

Copper is a trace element which is necessary for the metabolism, e.g. within the scope of the intracellular utilization of oxygen, the respiration. Copper is also necessary to keep the immune system working. Exposure to copper and copper oxide are to be found when using copper-containing welding and brazing fillers as well as when handling those joints. By suitable ventilation and production technological measures the tolerable MAC-values can be under-run easily.

Intoxications by inhalation of copper dust or smoke are very seldom. After inhalation of micro-distributed copper as dust or smoke in higher concentrations a metal smoke fever may develop. It is a reversible and short-term disease pattern with the symptoms of a common cold, which in the same way may be triggered by other metals (zinc, aluminum etc.).

After long-term inhalation of copper dust and smoke mucous membrane damage may develop in the nose. In the eye inflammation and damage of the cornea may develop by local impact of copper salts. In rare cases discoloration and chronic inflammation of the skin have been observed.

A chronic general intoxication by inhalation of copper at the workplace has not been observed until now, even when working for a long time in copper-rich surroundings.

8
Design Notes

8.1
Design for Material

Design for material requires the selection of the materials at an early stage of development. With the selection of materials a large part of the cost of manufacturing a component is defined. The most important for the selection of materials is the economic principle of "*sufficient functional efficiency at minimal expense*".

Important criteria for the selection of the materials are:
- Material properties
- Material cost
- Manufacturing properties
- Manufacturing cost
- Service life
- Weight

Copper and copper alloys have a high density and are expensive compared with other materials. This means: copper and copper alloys are only used in automotive engineering where properties are required that cannot be realized with lighter or cheaper materials.

The following properties are of central importance for the selection of copper or copper alloys:
- Excellent electrical and thermal conductivity
- Good sliding properties
- Good corrosion resistance

If it is only a question of high electrical or thermal conductivity, the best possible choice is one of the materials Cu-ETP, Cu-FRHC, Cu-OF, Cu-PHC or Cu-HCP, whereby the last two materials are to be preferred if welding or brazing is to be carried out. Accordingly for castings Cu-C-GS or -GP is to be selected. Requirements for weldability and brazeability of castings are to be agreed upon when ordering. If there are higher strength requirements, a compromise between strength and conductivity must be found and the low alloyed age-hardenable material CuCr1-C-GS or -GM may be considered.

Copper in the Automotive Industry. Hansjörg Lipowsky and Emin Arpaci

ISBN: 978-3-527-31769-1

Material cost depends upon the price of the base metal copper and of the alloying elements. Therefore the prices at a certain time will be given as an example. Prices of materials traded on the Metal Exchange (Cash buyer) are indeed fluctuating [69], the relations between different metals however (seen over several years) vary much less:

Material prices (approx.) early March 2000:

Cu: 1.80 USD/kg	Sn: 5.40 USD/kg	Al: 1.58 USD/kg
Pb: 0.46 USD/kg	Ni: 10.20 USD/kg	Zn: 1.09 USD/kg

Material prices (approx.) early March 2004:

Cu: 2.97 USD/kg	Sn: 7.10 USD/kg	Al: 1.64 USD/kg
Pb: 0.91 USD/kg	Ni: 14.50 USD/kg	Zn: 1.10 USD/kg

Material prices (approx.) at 27.07.2006 (1 EUR = 1.2688 USD):

Cu: 7.46 USD/kg	Sn: 8.24 USD/kg	Al: 2.45 USD/kg
Pb: 1.04 USD/kg	Ni: 27.04 USD/kg	Zn: 3.18 USD/kg

Copper is moderately priced in comparison with nickel even at 7.46 USD/kg (5.71 EUR/kg). The alloying element nickel is expensive and beryllium is extremely expensive. It follows from this comparison that copper–zinc and leaded copper–zinc alloys, particularly with high zinc content are cheap and nickel-bearing alloys are expensive. This, besides the cost for production, which is also influenced by the type of alloy, also needs to be taken into account for an estimate of the prices of semi-finished products, which means that the prices for semi-finished products do not increase as much as the price for copper and alloys. Nearly all of the copper and alloys for the automotive and supply industry is delivered in the form of semi-finished products.

About the processing properties of the wrought materials, it can be said that homogeneous materials such as un- and low-alloyed copper materials in a soft state, copper–zinc alloys with no more than 37% zinc and copper–tin alloys are excellently cold formable. It is not only possible to produce cheap semi-finished products (sheets, strips, tubes, etc.) of various dimensions of these materials, but they can also be excellently further cold worked, e.g. by bending, beading, upsetting, drawing, deep drawing, etc.

For example, connectors can be manufactured of the cheap material CuZn37 at low production cost by cold forming. Unfortunately, these connectors cannot always satisfy demands, be it because of higher temperatures or because of requirements concerning relaxation etc. This means that more expensive and less easily processable materials such as copper–tin alloys or special copper-based materials must be used. Strips of these materials are also characterized by good punchability, cold workability and narrow tolerances.

The special sliding materials of copper–tin and copper–zinc alloys with additional alloying elements (multi-alloys), e.g. CuSn8 and CuZn31Si1, which are very well suited for the production of wrapped bushings with thin walls and narrow tolerances, are also excellently cold formable.

The heterogeneous and normally leaded copper–zinc alloys are excellently hot formable, i.e. forgeable, materials. The material CuZn39Pb2 is the optimal solution if only cheap production of forgings of copper-based materials is required. This material is economical as regards material cost, can be forged at low cost and can also be machined (as free-machining alloy) very cheaply. However, there are also other copper materials such as Cu-ETP, Cu-PHC, Cu-HCP, CuCr1Zr and CuZn37Mn3Al2PbSi (for synchronizer rings) and copper–nickel and copper–aluminum alloys that are sufficiently forgeable. Copper–tin alloys cannot be hot formed and therefore cannot be forged.

Copper–aluminum alloys are highly corrosion resistant, high-strength, and very ductile materials. Deductions must however be made regarding formability of these high-strength, tough materials in comparison with the hot formability of leaded copper–zinc alloys. Semi-finished products of copper–aluminum alloys are harder to produce, as can be seen in the price.

Regarding machinability, the alloys CuZn36Pb3, CuZn39Pb2, CuZn40Pb2 and CuZn39Pb3, all free-machining, are better than any other. But the materials CuTeP, CuSP and CuPb1P are also very well machinable and besides nearly have the conductivity of unalloyed copper. Plain bearing bushings of the less cold formable heterogeneous materials, e.g. CuZn37Mn3Al2PbSi, CuSn12-C-GC or -GZ or CuSn7Pb15-C-GC or -GZ, are also easily produced by machining with narrow tolerances.

Apart from sliding purposes, cast materials are used relatively rarely in automotive engineering. For sliding purposes, whenever possible, centrifugally or continuous cast tubes or rods should be used. Both casting processes come very close to the final dimension in production of bearings, bushings, sliding rods, etc. Materials cast by these processes also show a favorable fine grain microstructure. For the same reason, castings produced in metal molds, e.g. chill castings, are to be preferred over castings from sand molds.

8.2 Design for Recycling

It should be ensured as early as possible in the design stage that at least the compact copper and copper alloy components can be dismantled easily by hand wherever possible. For example, cable harnesses, central electrics, starter, alternator, brake linings, heat exchanger, etc. should be as easy as possible to remove.

Small parts such as bearing bushings, connectors, contact springs, etc. cause no difficulties as copper materials can be relatively easily separated from shredder scrap due to their density and magnetic properties.

Recovery from components for electronic elements may be more problematic. Because of the different materials and because of some toxic substances attention should be paid to possibilities for simple dismantling of these components.

8.3 Limits of Application

Limitations for the application of copper and copper alloys are firstly visible from the density of these materials. As stated earlier, copper materials are only used in automotive engineering if properties are required that cannot be achieved with lighter or cheaper materials.

Copper materials are excellent low temperature materials, therefore no lower temperature limits are set. In contrast, however, when designing in the upper temperature range, creep limits need to be observed if the parts are in tension. Depending on the service lives copper–zinc alloys will start to creep under tension as soon as the temperature rises above 150 °C. Other copper materials such as copper, unalloyed or low alloyed, copper–tin, copper–zinc–tin, and copper–aluminum alloys should generally only be used up to 200 °C. Copper–nickel alloys are somewhat more heat resistant, meaning that CuNi10Fe1Mn can be used up to 250 °C and CuNi30Mn1Fe up to 300 °C. Copper materials can of course be used at higher temperatures if only slightly mechanically stressed and if the tendency to scaling does not restrict their application.

Copper–zinc alloys can fail due to stress corrosion cracking. They should therefore not be used in areas of the vehicle which are exposed to typical corrosive attack, e.g. by road salt, ammonia etc.

8.4 Tolerances and Machining Allowances

The relevant EN, ASTM and JIS specifications (product specifications) are listed in the Appendix “Standards and Specifications”. Regarding tolerances, the specifications for the various semi-finished product show:

- dimensions of strips and sheets
- dimensions of tubes
- dimensions for rods and profiles
- dimensions of wires
- dimensions and shapes of extruded profiles
- dimensions and shapes of hot stampings
- dimensions and shapes of open die forgings

For machining of rods on automatic lathes, specifically on machines with automatic rod feeds, so-called precision rods with bevelled and pointed ends have been developed. Products with a diameter or across flats size of 2 to 30 mm must be delivered with such ends (EN 12164) unless otherwise requested by the customer. The length tolerance for preferred lengths from 3000 to 4000 mm is at ± 50 mm. Machining allowances are normally not required.

Works producing semi-finished goods deliver ready-to-fit bushings in turned form or wrapped from strip. Bearing materials are standardized in ISO 4382-1, ISO 4382-2, ISO 4383 and ISO 4379.

The turned bushings production programme generally includes bushings with and without collars, half shells and turned parts in the areas:

- 9 to 200 mm ∅ for outside diameter
- 5 to 185 mm ∅ for inside diameter
- Maximum length up to 250 mm

It is possible to adhere strictly to the inside diameter tolerances. It is however important to use fine tolerances not generally but objectively. The average quality requirements specified by the standards normally suffice. Narrower tolerances can be supplied in special cases, if required. In the case of the highest demands the built-in bushings are to be reworked. Finer tolerances are specified for outside diameters and are easier to adhere to. Finish class N7 according to ISO 1302 is generally specified as surface quality. The eccentricity of bushings in the form of tube cut-offs is about ±5 to ±10%. The smooth surface from drawing has surface finishes of approx. $R_z=6$ µm. These bushings can also be supplied with machining allowances.

A typical example for bearings of tube cut-offs are spring eye bushings for HGVs.

Materials, shapes, dimensions and tolerances of wrapped bushings are detailed in the standards. The bore preferably has the tolerance H7. The inside diameter of the wrapped bushing shows tolerance H9 after pressing-in.

The general tolerances and machining allowances for copper castings are standardized in ISO 8062 where the tolerance limits are shown with corresponding drawings. Here, a distinction is also made between the tolerances of copper castings produced in sand molds, depending on whether they are hand- or machine molded. For castings, tolerated dimensions must be stated in the order as agreed between purchaser and foundry. The citation [21] contains tables with general tolerances and machining allowances.

The following general guide shows the roughness of copper castings:

- Sand casting: $R_a=18$ to 80 µm
- Shell mold casting: $R_a=$ 4 to 20 µm
- Precision casting: $R_a=$ 1 to 7 µm

9
Application

9.1
Areas of Application

Copper, in pure or alloy form is just as much an everyday material as it is a high-tech medium. Thanks to its versatility it is used in nearly all areas of everyday life. Large sectors of industry including transport, energy production and distribution, manufacturing and artisans use an incredibly wide range of copper materials.

About 40% of all copper applications are in the building sector, in electronic and sanitary installations alike, and about 38% in electromechanical applications. However, copper is also irreplaceable in new technologies. Both in new silicon chips or modern railway vehicles – copper can be found everywhere when a reliable high tech material is required.

About 9% of copper is used in the transport sector. The motor industry uses copper, above all, for electrical installations and electronics, such as wire harnesses and connectors. It also uses copper alloys for synchroniser rings and plain bearings, besides brake pipes, fans and other vehicle parts. Each vehicle contains from 18 to more than 28 kg of copper (the wire harness of a VW Phaeton alone contains more than 40 kg of copper) [11], that in future will be sourced more than ever as recycled material from secondary copper production.

9.2
Examples for Application

Without claiming completeness, Table 39 shows a list of examples for applications. The table lists the materials and also the semi-finished products.

Copper in the Automotive Industry. Hansjörg Lipowsky and Emin Arpaci

ISBN: 978-3-527-31769-1

Table 39 Examples for applications of copper and copper alloys in automotive engineering (materials and semi-finished products).

Examples for applications	Materials	Semi-products
Electric and electronic applications		
Wire harness	Cu-ETP/Cu-FRHC	wire, cord
Generator, alternator	Cu-ETP/Cu-FRHC	lacquered wire
Connector (contacts)	CuZn30, CuSn4, CuSn6, CuFe2, CuNiSi alloys	strips
Starter motor	Cu-ETP/Cu-FRHC	lacquered wire
Various electric motors (window, wipers)	Cu-ETP/Cu-FRHC	
Central electronics	CuZn37	sheets, strips
Ventilator, fan	Cu-ETP/Cu-FRHC	enamelled wire
Ignition solenoid	Cu-ETP/Cu-FRHC	
Various pumps (ABS, power lock)	Cu-ETP/Cu-FRHC	
Control units and relays	Cu-ETP/Cu-FRHC, CuSn4	strips
Contact springs, connectors	CuBe-, CuSn-, CuNiZn alloys, CuNi9Sn2, CuNi18Zn20	
Lead frames	CuFe2P, CuZn0, 5, CuSn0, 15, CuSn4	
Electrical resistors	Cu-Ni alloys	wire
Electrical contacts	Cu-PHC/Cu-HCP, Cu-Ni-alloys, CuNi30Fe2Mn2, CuNi9Sn2	microprofiles, plated
Plain bearings and applications for sliding movements		
Plain bearings, machined; sliding rods and elements	CuSn8, CuSn7Zn2Pb3-C-GC/GZ; CuZn31Si1; CuSn11Pb2-C-GC/GZ; CuSn7Pb15-C-GC/GZ; CuSn5Pb20-C	(hollow rods): continuous- and centrifugal castings
Bearing bushings, wrapped	CuSn8, CuZn31Si1	strips
Synchronizer rings	CuZn37Mn3Al2PbSi	hot stampings
Turned plain bearings in diggers	CuSn/ZnPb-C-GC/GZ	continuous- or centrifugal castings
Plain bearings in car steering systems	CuSn/ZnPb-C-GC/GZ	
Con rod bearings (steel backing shell)	CuSn11Pb2-C	composite casting
Cam shaft bearing, rocker arm bushing	CuSn8, composite steel casting	(hollow) rods
HGV spring eye bushings (tube cut-off)	CuSn8	tubes, hollow rods
Stub axle bushings	CuZn39Mn1AlPbSi	(hollow) rods
Valve guides	CuZn37Mn3Al2PbSi, CuSn-, CuNiSi alloys	

Table 39 (continued)

Examples for applications	Materials	Semi-products
Other applications in automotive engineering		
Water and/or oil coolers Vehicle radiators	Cu-DHP, CuZn28, CuZn33, CuZn36	foils, strips, tubes *Cu*proBraze®
Automatic and contour turned parts, e.g., carburettor nozzles, cylinder locks	CuZn39Pb2	rods, profiles
Window winder mechanics	CuZn39Pb2	
Shift forks, forged and cast	CuZn39Mn1AlPbSi, CuZn37Mn3Al2PbSi, CuAl10Ni5Fe4, CuAl10Fe5Ni5-C-GM	hot stampings, chill casting
Ball bearing cages	CuZn39Pb1Al-C-GC/GZ, CuZn36	continuous- or centrifugal casting: tubes
Hydraulic brake pipes	CuNi10Fe1Mn	tubes
Intercooler, charge air cooler	CuNi10Fe1Mn	
Pinion for ignition distributors	CuZn37Mn3Al2PbSi	rods
Heat exchangers for air conditioning	Cu-DHP, CuZn30, CuZn36	foils, strips, tubes
Fuel, oil pipes	Cu-DHP	tubes
Electrode materials (resistance welding)	CuCr1Zr, CuBe-, CuNi alloys	rods
Welding electrode holder	CuCr1-C	casting

10
Usage Properties

10.1
Corrosion Behavior

10.1.1
Basics of Copper Corrosion [1, 14, 24, 70]

The behavior of pure copper is the basis for corrosion behavior of copper alloys. The corrosion behavior of pure copper will therefore be treated first.

The *corrosion* of metals is nearly always caused *electrochemically*. This means, that normally only a liquid or moist medium, an electrolyte, can cause corrosion – except hot gas corrosion. The corrosion of metals is due to the fact that most work metals are ignoble and strive for a return to the natural state, from which they are produced by consumption of energy. The electrochemical series gives the first indication of durability. Corrosion resistance is however influenced by further factors such as concentration, temperature and flow speed of the medium by which the electrochemical potential is displaced and formation of passivating layers. This leads to considerable dispersion of the electrochemical series, referring respectively to the aggressive medium.

If two different metals are immersed in an electrolyte, they will produce a galvanic cell. Between both elements, voltage can be measured corresponding to nearly the difference of the normal potential of both metals. If energy is taken from the cell, this will take place at the cost of the less noble metal (anode), which dissolves dependent on the current drain. The extreme case is the short circuit of the element through direct contact of both *metals* (*local element*). This term is of decisive importance for the explanation of corrosion procedures and for establishing the cause. It is important to know that also two neighboring points of the same metal which differ in their surface states can form a local element. Such local differences can result from inclusion of foreign metal, heterogeneous microstructure, various conversion layers, different ventilation or concentration due to deposits etc. This can lead to potential differences of up to 0.1 V. With this in mind, assessment of the practical, normal electromotive series in a real application case is not sufficient to evaluate the corrosion risk, but gives an initial indication of possible corrosion dangers.

Copper in the Automotive Industry. Hansjörg Lipowsky and Emin Arpaci

ISBN: 978-3-527-31769-1

Copper is the most noble work metal and its position in the electrochemical series above hydrogen explains its good corrosion resistance. Based on its more noble potential in solutions with hydrogen ions, copper is not affected by water, aqueous solutions or non-oxidizing acids in the absence of oxidizing agents. Copper is also not affected by weak media, e.g. in the atmosphere or in oxygen-containing water as it is capable of forming protective layers as explained in the previous paragraph.

Corrosion speed generally rises with *temperature* in the case of copper materials as the course of electrochemical reactions is accelerated by increasing temperature. On the other hand, the corrosion speed of copper-based materials can also decrease with rising temperature as corrosion depends on oxygen content in aqueous media and it is known that oxygen solubility in water diminishes as the temperature rises.

It therefore follows that pure copper (e.g. Cu-DHP) is resistant against industrial water as used for example in oil coolers and against drinking water. Tubes of Cu-DHP are used to a large extent for hot and cold drinking water pipes. In water, copper forms oxide conversion layers whereby free carbon dioxide can slow down the formation of those layers. Some copper is initially dissolved in such acidic waters. This however does not lead to corrosion damage. Local elements, e.g. because of deposits (sand or rust), are normally the cause of pitting in water pipes.

Copper may be affected if the medium contains oxygen or oxidants or is itself oxidizing, as for example nitric acid. Corrosion resistance of copper is also endangered if the copper forms complex salts with the aggressive agent which impair the protective layer. To these aggressive agents belong in particular ammonia, ammonium-containing solutions, cyanides and concentrated haloid acids.

10.1.2
Types of Corrosion

The least harmful form, in most cases, of corrosion of copper materials is the uniform loss of material at the surface. Here, the metal is evenly dissolved over the entire wetted surface, often at a constant and low speed. The process can often be halted by protection and conversion coatings (e.g. formation of an oxide layer). This process is only then critical if the component can be weakened too much due to dissolution of the metal during its service life.

Pitting corrosion affects very narrowly defined areas of the surface and penetrates deep into the metal. The corrosion is caused electrochemically and is almost always due to a variety of deposits by which the uniform access of the oxygen in the corrosive medium is interrupted. Such deposits can be solid dirt particles, grease residue, sand grains or corrosion products leading to local disruption of the otherwise protective conversion coating that is being formed.

Also at the borderline between liquid medium, air and metal, this can lead to a very fast anodic partial reaction (metal dissolution). Pitting corrosion takes place mainly in tubes. Surface corrosion takes place near the borderline and is to be expected in case of standstill times or partial emptying of containers or appliances.

Selective corrosion means specific corrosion of a number of alloys by which one alloy element is dissolved and leaves the other in the form of a sponge (e.g. dezincification of brass). Corrosion can be sheet-like or plug-like in nature and affect the strength of the component or lead to a break-through in the wall.

Contact corrosion is normally unavoidable or appears as the result of design mistakes if conductively connected metals of different potential are exposed to a corrosion medium (electrolyte). Even without conductive assembly of metals with electromotive potential difference, corrosion may occur if local elements can be formed with the help of a flowing corrosive medium. This is, for example, the case if a liquid initially flows over the surface of copper materials and then over the surfaces of less noble metals, e.g. steel, zinc or aluminum. Dissolved traces of copper precipitate on the surface of the steel, form a local element and lead to corrosion. A demonstrative example is rainwater which flows over the copper sheets of a chimney down into the eaves gutter of zinc or galvanized steel which will then be dissolved. In closed systems this type of corrosion can be minimized or prevented by adding an inhibitor to the liquid medium.

The potentials of the different copper alloys are so close to each other that they can be interconnected without danger. Stainless steel normally also gets along quite well with copper. Nearly all the other work metals and alloys of automotive engineering become sacrificial anodes in contact with copper materials and may be attacked rather quickly.

Inter-crystalline corrosion may also occur in rare cases. The grain boundaries of metallic materials, in particular with coarse grain structures, are always weak spots in which foreign atoms and contaminations become enriched so that they are less noble than the grains themselves. Corrosion then takes place preferentially at the grain boundaries and advances along the grain boundaries into the inner structure.

A form of corrosion also known to affect copper alloys is stress corrosion cracking (SCC). This takes the form of sudden brittle fracture of the component or the sudden occurrence of cracks mostly without any corrosive attack being visible in advance. Stress corrosion cracking can only occur if three conditions are fulfilled simultaneously:

- The material must be susceptible to stress corrosion cracking due to its composition.
- There must be residual or service tensile stress in the surface zone of the material.
- The material must be exposed to certain specific corrosion agents.

Erosion closely resembles the damage caused by corrosion even though this is brought about by mechanical stress of flowing or running steam or liquids. Erosion is generally accompanied by corrosion. Erosion-corrosion limits the acceptable flow speeds of a medium in tubes. Some alloys are more resistant than copper.

Closely related to erosion is *cavitation*. It is caused by imploding bubbles of vapor of a flowing liquid which form in regions of high flow speed and therefore low pressure of the liquid. The damage mechanism is in fact a mechanical one as the stress induced by the implosions leads to fatigue of the material. This is frequently accompanied by corrosion. This is probably why cavitation is mostly summarized under "corrosion" or "erosion". As the main damage mechanism is fatigue, selection of a material with higher strength will extend service life. Higher corrosion resistance mostly helps.

The most insidious type of corrosion is *corrosion fatigue* as it also attacks materials which are widely taken as corrosion resistant. The fatigue strength may be drastically reduced under the influence of simultaneous cyclic loading and corrosion attack. Copper materials are also prone to corrosion fatigue. Some alloys behave better than pure unalloyed copper.

Section 6.2 deals with the fatigue strength of copper materials. As stated there, most fatigue fractures start at the surface. Extrusions and intrusions that appear on all cyclically mechanically loaded metallic materials disrupt passivating (protective) layers and leave the base metal bare to the corrosive attack. These bare regions take the role of a sacrificing anode and are therefore further preferentially attacked. This again speeds up progress of the fracture. So corrosion fatigue is an interaction between electrochemical actions and cyclic mechanical stress whereby both failure mechanisms are enhanced. The cracks are without plastic deformation and run in a fairly straight line vertical to the main direction of tensile stress.

10.1.3
Corrosion Behavior of Copper Alloys [70]

Low alloyed copper materials behave essentially like unalloyed copper as regards corrosion resistance. The corrosion behavior of copper alloys can be significantly influenced by alloying elements and – in many cases – improved. Corrosion behavior is slightly improved by the alloying elements Si, Cr, etc. Aluminum for example is rather ignoble but in fact quite corrosion resistant. It forms spontaneously a naturally stable passivating conversion layer of aluminum oxide that protects the metal. Aluminum transfers this property to copper materials. This is why also aluminum-bearing copper alloys such as copper-aluminum and aluminum-bearing brass are more resistant than pure copper in several corrosive media.

Corrosion resistance of *copper–zinc alloys (brass)* is determined to a large extent by the copper content. The high copper-containing α-brass behaves similarly to pure unalloyed copper. Corrosion behavior can be considerably improved by adding other elements to the copper–zinc alloys (special brass). An example of this is the aluminum-bearing copper–zinc alloy CuZn20Al2As, which is highly resistant against seawater.

The heterogeneous *copper–zinc alloys* with zinc contents of 30% or more, the so-called $(\alpha+\beta)$ brass types are also characterized by good corrosion resistance but are inferior to the corrosion resistance of unalloyed copper. With copper–

zinc alloys, two special types of corrosion must be given special consideration to prevent corrosion failure: Possible selective corrosion (dezincification) and stress corrosion cracking (SCC) (see Section 10.1.2).

While the share of β-mixed crystals in *copper–zinc alloys* improve some characteristics, e.g. strength and machining properties, the β-mixed crystals adversely affect the corrosion resistance. The susceptibility to selective corrosion is increased. Added to this is the fact that only α-mixed crystals can be protected by corrosion inhibitors, e.g. 0.02 to 0.035% arsenic against dezincification. Dezincification occurs in two forms, mainly dependent on the carbonate water hardness in waters with high chloride content. Distinction is made between a so-called layer dezincification in the form of uniform surface corrosion and plug dezincification in which cavities are formed and filled with corrosion products. For applications in which selective corrosion must be expected, mills producing semi-finished products supply so-called "dezincification-resistant copper–zinc alloys", which are inhibited against dezincification and are specially heat treated.

Furthermore, *copper–zinc alloys* with more than 15% zinc tend towards *stress corrosion cracking (SCC)* if under service and/or residual tensile stress and simultaneously under the influence of certain corrosive media such as ammonia, amines, sal ammoniac and sulfur dioxide. With this kind of corrosion, the component shows widely ramified trans-crystalline and inter-crystalline cracks. Just slight traces of ammonia or sal ammoniac in humid or oxygen-containing atmospheres, as appear in e.g. agricultural areas, are sufficient. For protection measures see Section 10.1.5.

Nickel–silver alloys are distinct from brass by the nickel alloy content, which significantly improves corrosion resistance compared with brass. Weathering resistance has been specially improved while the tendency to local corrosion and stress corrosion cracking has been significantly reduced.

Copper–tin and copper–tin–zinc alloys (tin bronze and red brass) together with the copper–aluminum and copper–nickel alloys are among the most corrosion resistant copper materials as they cover themselves with firmly adhesive, tight protective layers. Copper–tin alloys are only slightly endangered by pitting corrosion and, in contrast to copper–zinc alloys, immune from selective corrosion and stress corrosion cracking.

Tin-bearing copper alloys are excellently corrosion resistant in the atmosphere, in aqueous, in weakly acidic and in weakly alkaline media and in non-oxidizing acids. They are seawater resistant and lead contents improve their resistance to weak acids.

The excellent corrosion resistance of *copper–aluminum alloys* is due to the fact that the already good resistance of copper is increased by a layer of Al_2O_3 on the surface which is tight, self-healing and sticks firmly to the surface. Copper–aluminum alloys are seawater resistant and have established themselves for a number of corrosive media too aggressive for other copper materials. These range from weak acids to weak alkaline salt solutions, organic, reducing or weakly oxidizing mineral acids. The tendency to selective corrosion (de-aluminification) is very slight. These alloys are immune from stress corrosion cracking.

Copper–nickel alloys are among the most resistant copper materials. They are resistant against humidity, seawater, non-oxidizing – also organic – acids, alkaline- and salt solutions and dry gases such as oxygen, chlorine, hydrogen chloride, hydrogen fluoride, sulfur dioxide and carbon dioxide. These materials are immune to stress corrosion cracking and the tendency to selective corrosion and pitting corrosion is very low. Copper–nickel alloys are superior to all other copper materials with regard to erosion resistance. Erosion resistance increases with nickel content and the highest flow speeds are therefore permitted for tubes of CuNi30Mn1Fe.

10.1.4 Corrosion in Gases

In clean, dry atmosphere bare metallic copper covers itself with a thin layer of copper(I) oxide. Although this protective layer can be invisible, it increases the corrosion resistance of the already noble copper. This is why copper is highly corrosion resistant in the atmosphere.

If copper is exposed to an industrial atmosphere without its protective layer being fully developed, the red coloring on its surface will first change, due to oxide layer formation, from red brown to dark brown and gray black. Color formation depends on the state of the copper's surface and humidity and the sulfur content in the atmosphere. This protective layer normally becomes a layer of green patina in time. This consists of a mixture of water-insoluble, alkaline copper salts (sulfate, carbonate and also chloride, if near the sea). Ammonia and hydrogen sulfide impede the formation of the protective layer.

Copper develops a thin tarnish film when heated in air to approx. 125 °C. Scale will not commence until a temperature of approx. 250 °C is reached whereupon the red layer of Cu_2O is covered with a black coating of CuO.

Hydrogen and H_2-containing gases cause hydrogen-embrittling in oxygen-containing copper types such as Cu-ETP or Cu-FRHC at high temperatures above 500 °C (hydrogen disease, hydrogen embrittlement). Hydrogen diffuses into the copper and reacts there with Cu_2O to form water vapor. Water vapor cannot diffuse out of the copper and because of its high pressure at high temperatures disrupts the microstructure. Oxygen-free copper, e.g. Cu-DHP, is insensitive to hydrogen embrittlement. The carrying out of hydrogen embrittlement tests on deoxidised and oxygen-free highly conductive copper is laid down in EN ISO 2626 (Copper-hydrogen-embrittlement tests).

10.1.5 Corrosion Protection

Additional corrosion protection for copper-based materials is normally not necessary.

If there is a high risk of selective corrosion (dezincification) for components of copper–zinc alloys, as the parts are exposed to high chloride content, e.g. by de-icing salt, in a humid atmosphere, it is recommended to use inhibited materials, e.g. copper–zinc alloys inhibited with small amounts of arsenic. The corrosion test is to be carried out according to EN ISO 6509 (Corrosion of metals and alloys – Determining the dezincification resistance of copper–zinc alloys).

It is recommended to thermally stress-relieve the materials if brass components with zinc content over 15% are under residual and/or service tensile stress. It cannot be excluded that vehicle components, e.g. in rural regions, come into contact with traces of ammonia from the atmosphere. These parts are thus at risk of stress corrosion cracking. Stress-relieving considerably reduces corrosion risk as the occurrence of stress corrosion cracking is crucially dependent on the level of tensile stress. The heat treatment of stress-relieving is treated in detail in Section 7.3.3. The corrosion test is carried out according to EN ISO 196 "Copper and copper casting alloys – Locating residual tension – Mercury(I) nitrate test" and EN 14977 "Copper and copper alloys – Locating tensile stress – 5% ammonia test" and/or ISO 6957 "Copper alloys – Ammonia test for stress corrosion resistance". The mercury nitrate test is highly sensitive and indicates the presence of residual stress by cracking open. Tests nowadays prefer the use of ammonia due to environmental concerns.

Electroplatings improve resistance to dezincification and stress corrosion cracking. These do not however offer absolute protection as the layers are too thin and mostly porous. For these reasons and because of the expense it is normally done without this protection. Only sensitive components for electronic applications are tin-plated or protected with noble metal platings.

Corrosion inhibitors that are added in lower concentrations play a major role in closed cooling and hydraulic systems. Complex-building agents of the benzotriazol type have become established for copper-based materials. These inhibitors form tight bonds with the dissolving copper ions. When using these inhibitors, it should always be checked whether there are other materials in the loop and if the inhibitors could cause undesirable reactions with these.

10.2 Strips for Electrical Engineering and Electronics

As these strips, besides the sliding elements, constitute the largest share of applications in vehicle manufacture, the usage properties will be treated in more detail here.

Strips have for a long time been used as spring strips and for connectors. New is the application as lead frames for integrated circuits. Low alloyed copper materials are being used more and more for these applications.

The following criteria are used for the selection of materials for spring strips:

- *Elastic behavior*: Young's modulus E and spring bending limit B (FB) determine the characteristics and power of the spring (see Section 6.2.1).
- *Cyclic load capacity*: The load limit is determined by the fatigue properties under alternating flexural stress (see Section 6.2.6).

- *Stress relaxation tendency*: Relaxation gives a measure for the setting of the springs at static elastic long-term load (see Section 6.2.3).
- *Electrical conductivity*: Electrical conductivity decides about the selection for power conducting springs (see Section 6.1).
- *Magnetic properties*: Permeability plays a role for measuring devices (details are given in Ref. [71]).
- *Chemical behavior*: Corrosion- and tarnishing resistance and soldering properties are of great importance for storage, processing and functional reliability (see Sections 7.2.1 and 10.1).
- *Processability*: Punching properties and bendability are important but so are weldability and suitability for surface treatment (see Section 7.4).

Spring strips are generally used in the hard or cold-worked state. The selection of materials is thus determined by the requirements and the specific properties of the materials. Tables 24 to 27 contain a selection of materials for spring strips with their mechanical properties.

Regarding the elastic behavior, distinction is made between two material groups: Materials with a Young's modulus of 110 to 125 kN/mm^2, e.g. CuSn5, CuSn6, CuSn8, CuFe2P and CuZn23Al3Co, and materials with a Young's modulus of 135–140 kN/mm^2, e.g. CuNi9Sn2, CuNi18Zn20, CuBe2 and CuCo2Be.

For a required spring deflection or spring strength, the material is selected depending on the ratio B (R_{FB} or FB) to E. Spring materials have spring bending limits of 300 to 800 N/mm^2. For materials for cold rolled spring strips, the spring bending limit is considerably increased by tempering at temperatures of 200 to 300 °C. Firstly, spring strips are thereby thermally stress-relieved, and secondly, temper-hardening is achieved with brass, nickel silver and tin bronze. The values for the Young's modulus and spring bending limit depend on whether they are measured parallel or transverse to the rolling direction because of the texture produced by cold rolling.

The fatigue properties under alternating flexural stress as a measure of cyclic load-bearing capacity of power conducting springs have already been explained in Section 6.2.6. Figure 16 shows the fatigue strengths of a number of spring materials. The bending fatigue behavior is described by the so-called "Wöhler curve". The curve shows that higher stress amplitudes are tolerated only with a lower number of cycles-to-failure. The fatigue strength under cyclic stresses depends on the tensile strength, the thickness of the strip and the grain size. Finer grain microstructures have a higher fatigue strength than coarse grain.

Relaxation has already been treated in detail in Section 6.2.3. The gradual loss of elastic spring tension in the course of time under the influence of elastic load is characterized by the term "relaxation". Relaxation increases with the temperature and rises with initial spring tension but depends on the material as shown by Fig. 17. Tension will drop faster with copper–zinc alloys while nickel-bearing materials have only a slight relaxation tendency and tin bronze lies somewhere in between. Work-hardened materials behave similar to age-hardened materials.

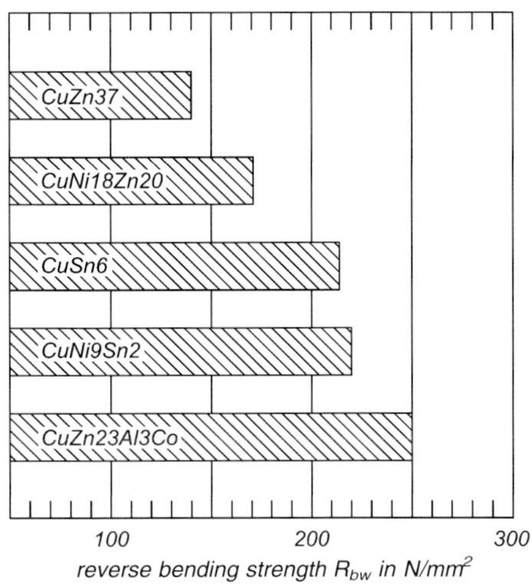

Fig. 16 Fatigue strength under alternating flexural stress of a number of spring materials in a cold-worked state at 10^8 load cycles (measured at approx. 1 mm thick strip samples, parallel to the rolling direction).

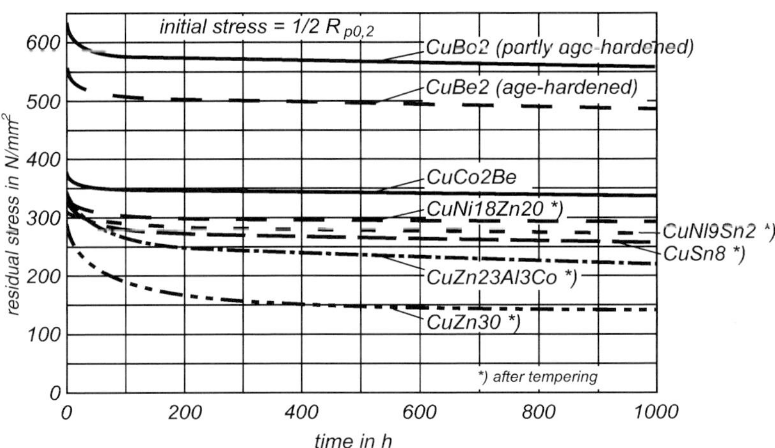

Fig. 17 Relaxation behavior of various copper materials at 150 °C.

The selected material determines electrical conductivity and spring bending limit in the case of power conducting springs. It must be borne in mind that materials with high strength but low conductivity can heat up considerably, which can also reduce the spring bending limit. Table 21 contains details on electrical conductivity and Tables 24 and 27 deal with the spring bending limit.

Magnetic properties are important for measuring devices. Copper materials have favorable magnetic properties as they are, in principle, diamagnetic and paramagnetic [71]. Permeability of copper materials is near 1.

The good corrosion resistance of the copper alloys has already been treated in detail in Section 10.1. The susceptibility of copper–zinc alloys to stress corrosion cracking in a humid atmosphere is to be kept in mind when using spring materials. Spring materials CuZn37 and CuZn30 are at particular risk, whereby susceptibility can be significantly reduced by stress-relieving (see Section 7.3.3). Contamination from a humid atmosphere can lead to the formation of tarnish films on power conducting spring surfaces that can impair their functioning. The material CuNi9Sn2 is highly resistant to tarnish films and other copper-based spring materials are tin-plated or plated with noble metals for such applications. Tarnish films can also impair solderability so that for long-term storage the use of tin-plated strips is recommended.

All copper materials can be excellently electroplated. Zinc evaporation can disrupt welding and so zinc-free materials should be used for such operations.

An especially important property for spring strips is their bendability as connectors and many spring elements are produced by folding or bending. Details on the smallest possible bending radii are contained in Tables 24 to 27. It is to be borne in mind that bendability is dependent on direction (parallel or transverse to the rolling direction). Bendability decreases with increasing tensile strength (hardness). Tin bronze and the material CuZn23Al3Co are particularly favorable regarding bendability.

Following development of the semi-conductor elements, a technology became established for assembling elements on copper material precision cutting strips and then encapsulating them. It was only possible with the help of the lead frames to achieve automated mass production for semiconductor elements and integrated circuits. The following material properties are required for the selection of strips for lead frames:

- High thermal and electrical conductivity
- Good punching properties for the strips (no burr)
- Good temperature and tempering resistance
- Good solderability and electroplating properties
- Good corrosion resistance

Table 40 shows the most common lead frame materials with their physical properties.

Table 40 Physical properties of some copper-based lead frame materials.

Symbol	Density (g/cm^3)	Electrical conductivity (MS/m)	Electrical conductivity (% IACS)	Heat conductivity (W/m K)	Coefficient of thermal expansion (10^{-6} K^{-1}) (RT to 300 °C)
CuSn0.15	8.9	≥50	≥86	360	17.3
CuZn0.5	8.9	52	90	350	17.7
CuFe2P	8.8	35	60	260	16.3 [a)]

a) Dependent on the measuring process and specimen condition other values may occur, e.g., 17.4 or 17.6.

The required properties are partly identical with those for power conducting springs. Therefore, only the differing requirements will be treated here. The high demands on thermal and electrical conductivity are, among other things, based on the fact that with power semiconductors relatively high power loss in the form of heat is to be dissipated. The electrical conductivity of the materials CuZn0.5 and CuSn0.15 at over 85% IACS and accordingly heat conductivity are very high. The highest conductivity possible is required for surface mounted devices (SMD components). The most important lead frame material CuFe2P indeed has lower conductivity values (60% IACS) but is stronger than the other two materials. The tempering resistance or softening temperature of the lead frame materials CuZn0.5 and CuSn0.15 is so high that the temperature of 150 °C arising at thermosonic bonds is well below this. The material CuFe2P is even more favourable in this regard.

10.3 Plain Bearings [14, 72]

Strips for electrical and electronic applications represent the highest portion of the copper-based materials in automotive engineering but plain bearings of copper material are also used to a large extent in this field. Compared with roller bearings, plain bearings are advantageous for some applications as shown by the comparison in Table 41.

Plain bearings must be used if roller bearings are not recommended because of low speeds or small glide paths or high surface pressure. Plain bearings also

Table 41 Comparison of the properties of roller- and plain bearings.

Property		Roller bearings [a]	Plain bearing [a]
Construction length		⇑	⇓
Required space for outside diameter		⇓	⇑
Friction when warming up		⇑	⇓
Friction at middle speed		⇔	⇔
Friction at high speed		⇓	⇑
Expenditure for lubrication		⇑	⇓
Behavior on influence of dirt		⇓	⇑
Behavior when receiving shocks and jolts		⇓	⇑
Noise and oscillations		⇓	⇑
Cost of design and calculation		⇑	⇓
Costs of split bearings		⇓	⇑
Cost of replacement under normal circumstances		⇔	⇔
Cost of replacement under exceptional circumstances		⇔	⇔
Service life	Plain bearings with liquid friction	⇓	⇑
	Plain bearing with deficient lubrication	⇔	⇔

a) Change: ⇑ = superior; ⇓ = inferior, ⇔ = equal.

need considerably less space and those with full hydrodynamic lubrication have almost unlimited service lives. The load bearing capacity of roller bearings decreases quickly with increasing speed.

Copper materials have been showing their worth as materials for plain bearings for years. They excel by a combination of various properties, which are to be adapted to the respective application and the sliding partner (shaft).

- High pressure load-bearing capacity
- High wear resistance
- Good embeddability for foreign matter and high ductility
- Sufficient fatigue strength
- Good corrosion resistance
- Low tendency to weld with the sliding partner
- Good emergency running properties
- Low coefficient of sliding friction
- High heat transfer
- Good machinability

The ductility of the material, its adaptability and a limited embeddability for foreign bodies in the sliding surface can be influenced by the choice of material and condition. The choice of hardness ranges from HB=45 with the material CuSn5Pb20-C-GS to HB=200 with CuAl10Ni5Fe4. Table 42 shows a number of physical properties for a selection of sliding materials commonly used in automotive engineering.

Table 43 contains the mechanical properties of the same sliding materials. Further values such as fatigue strength, etc., for casting materials are contained in Ref. [21].

A minimal tendency to welding (cold welding, galling) is determined by properties such as chemical bonding of the sliding partners, mutual insolubility and level of the adhesive forces. Estimation of the values confirms the old rule that in plain bearings materials as different as possible, e.g. tin-bronze and steel, should be combined with one another.

The energy to overcome the adhesive forces, i.e. energy for deforming and levelling the surfaces, is to a large extent to be dissipated as heat. Local and short-term temperatures of 1000 °C because of the friction forces can thus occur and lead to local welding and heavy wear. Materials with high heat conductivity show clear advantages.

Plain bearings of copper materials can be divided into:

- Solid plain bearing of cast or wrought materials
- Powder metallurgically produced sintered bearings
- Composite plain bearings

Solid plain bearings have a heterogeneous dendritic microstructure and normally also contain lead as an insoluble alloying element. The good emergency running properties are largely determined by the lead content. Cast materials produced by continuous- or centrifugal casting are preferred. These casting pro-

Table 42 Physical properties and elevated temperature hardness of a number of sliding materials used in automotive engineering.

Material code:	Hardness HB 10 at						Heat conduktivity (W/m/K)	Coefficient of thermal expansion (10^{-6} K^{-1})
	RT	50 °C	100 °C	150 °C	200 °C	400 °C		
CuSn7Zn2Pb3-C-GC/GZ	70–100	70	68	64	60	–	59	18
CuSn7Pb15-C-GC/GZ	65–75	65	63	61	60	–	47	18
CuSn5Pb20-C-GC/GZ	50–65	58	56	54	52	–	59	19
CuSn11Pb2-C-GC/GZ	80–140	78	75	–	70	–	54	18
CuSn8	80–>160	100	99	99	98	–	59	17
CuZn37Mn3Al2PbSi	145–>170	140	125	–	100	20	63	20
CuZn31Si1	100–>160	110	100	–	85	–	67	18
CuAl10Ni5Fe4	170–220	178	172	171	171	127	27	16

Table 43 Mechanical properties of a number of sliding materials in automotive engineering.

Material code	Young's modulus E (kN/mm²)	Tensile strength R_m (N/mm²)	Elongation A_5 (%)	0.2% proof strength $R_{P0,2}$ (N/mm²)	Hardness HB	Compression strength R_{dB} a) (N/mm²)
CuSn7Zn2Pb3-C-GC/GZ	93–115	270	16/13	120/130	70/75	880
CuSn7Pb15-C-GC/GZ	75–82	220	8	110	65	685
CuSn5Pb20-C-GC/GZ	74–81	165	6	90	50	390
CuSn11Pb2-C-GC/GZ	90–110	280	7/5	140/150	85/90	1080
CuSn8	108–115	450	30	245	80/160	–
CuZn37Mn3Al2PbSi	98–117	590	18	275	140/170	1080
CuZn31Si1	98–108	440	25	245	100/160	–
CuAl10Ni5Fe4	114–118	640	10	340	180	1180

a) Compression strength is defined as the stress at which the first crack or fracture is measured in the compression test or an overall total compressive strain of 50% is achieved.

cesses are very economical for the production of sliding materials. Due to their rapid solidification, these materials also have a uniform fine microstructure with improved mechanical and thus also improved sliding properties.

Solid bushings or sliding rods of wrought copper materials have finer grain microstructures than cast materials. Strengths can also be adapted to requirements by work hardening. Not only can plain bearings be produced from wrought materials by machining, e.g. by turning of a drawn tube, but wrapped bushings can be produced from strips extremely economically. They can be mounted in a space-saving manner but strict demands on shape and position accuracy cannot be fulfilled.

Turned bushings however fulfil the highest demands on accuracy and tolerances. They are certainly more expensive than wrapped bushings due to their

Table 44 Characteristics of the load-bearing capacity for continuous or centrifugal cast solid plain bearings or wrought copper materials.

Material code:	Load-bearing capacity (N/mm^2)	
	Oscillating bearing	Rotating shaft (v>2 m/s)
Cast materials		
CuSn7Zn2Pb3-C-GC/GZ	50	30
CuSn11Pb2-C-GC/GZ	80	40
CuSn7Pb15-C-GC/GZ	40	30
Wrought materials		
CuSn8	100	40
CuZn31Si1	90	40
CuZn37Mn3Al2PbSi	100	40

expensive manufacture. Table 44 shows a number of reference values for load-bearing capacities of copper castings and wrought copper materials. Table 45 contains notes on sliding properties and possibilities for application.

For solid bearings, the sliding partners should be selected according to the load. It is generally recommended for the sliding partner (generally the shaft) to be about 60 to 80 HB harder than the sliding material. As regards running-in behavior, it can be said that plain bearings of leaded materials such as tin–lead bronze or red brass run in on putting into operation by plastic deformation. Harder tin bronze, cast and wrought materials run in by plastic deformation and abrasion and copper–zinc or copper–aluminum alloys only run in by abrasion. This must be considered when selecting lubricants and in the design of lubrication grooves.

Besides the production of plain bearings of semi-finished products, the production of sintered plain bearings is of technical importance. The most important materials for sintered plain bearings are laid down in ISO 5755 (Sintered materials – Specifications). By sintering, sliding elements are produced with a pore volume up to 25%, which – subsequently impregnated with solid or liquid lubricant – have self-lubricating properties. These sliding elements are maintenance-free for a limited period and independent of the supply of lubricants. It must be kept in mind in the case of machining that the sliding surfaces should not be machined as the pores may be clogged hereby. Sintered bronze with 8.5 to 11.0% tin (without graphite or with 0.5 to 2.0% graphite content) is nearly exclusively used as the material for sintered bearings. Further details on these maintenance-free small bearings are contained in Refs. [73, 74].

The production of wrapped bushings is the most economic way to produce plain bearings. Composites have been developed to facilitate the production of wrapped bushings of highly leaded materials and at the same time to increase the fatigue strength of the bushings. For these composites, a sliding layer is per-

Table 45 Instructions for the use of continuous or centrifugal cast solid plain bearings, wrought copper or composite cast materials.

Material code	Special properties	Application
Cast materials		
CuSn7Zn2Pb3-C-GC/GZ	sliding material for non-critical applications under moderate load and good lubrication	good emergency running properties, guide elements, platform gate fasteners
CuSn12Ni2-C-GC/GZ	insensitive to pitting as it is lead-free	bevel gears, worm gear crowns
CuSn11Pb2-C-GC/GZ	ductile, hard material with good sliding properties and wear resistance for high stress and high speed	condition: well-aligned mounting with good lubrication; hardened shafts; toggle bearings
CuSn7Pb15-C-GC/GZ	soft sliding material for moderate loads and moderate to high speeds	lead increases the emergency running properties; is insensitive to edge pressure
CuSn5Pb20-C-GC/GZ	soft sliding material for moderate loads, moderate to high speeds; composite casting material	lead increases the emergency running properties; is insensitive to edge pressure
Wrought materials		
CuSn8	hard, ductile sliding material for high loads and high sliding speeds, good lubrication	condition: well-aligned mounting and hardened shafts; conrod bushings, rocker bearing
CuZn31Si1	for high load, moderate to high speed, good lubrication	condition: well-aligned mounting and hardened shafts; piston bosses, front axle spindle bearing
CuZn37Mn3Al2PbSi	high wear-resistant sliding material suitable for moderate lubrication	condition: well-aligned mounting and hardened shafts; switch pinion bearing, extension arm for diggers
CuAl10Ni5Fe4	very hard alloy, requires good lubrication, shock resistant, highly resilient	condition: well-aligned mounting and hardened shafts; toggle bearing

manently joined to a steel strip by casting, sintering or roll-cladding. With this composite, a material is created that unites the good properties of copper–tin alloys with the strength and low thermal expansion of steel. These processes enable the production of strips of composite materials, that are then used to produce wrapped bushings. In this way, the mechanical strength of bronze increases as the thickness of the sliding layer decreases. It is generally for this reason that layer thicknesses of just a few tenths of a millimeter are usual for composites for sliding purposes.

The cast-plated strips with copper–tin–lead sliding layers, e.g. CuSn5Pb20-C-GS, are mainly used for internal combustion engines. High-performance bearings are also given an additional thin electroplated sliding layer, e.g. of a lead–tin alloy, hardened with copper. Specifically higher loaded conrod bearings often consist of a steel back with a layer of CuPb24Sn4, a nickel wall above this and an electroplated sliding layer of PbSn10Cu2. This material can be used up to a specific bearing stress of approx. 65 N/mm^2. PVD layers of aluminum-, tin- or aluminium–lead are additionally used for special applications.

Besides the metallic three-layer composites, there are also those with a wide variety of synthetic sliding layers. Such plain bearings show good heat conductivity with their thin synthetic layers and the tight cottering with the tin bronze framework and have the good sliding properties of synthetics, both for lubricated applications and dry running.

10.4 Reliability

The reliability of materials used in automotive engineering is decided by various factors ensuring long-term operation without malfunction.

Although copper materials are seldom purely mechanically stressed, it should be pointed out that the properties of copper do not change due to something like aging during service life. Above certain temperatures behavior under long-term load, and in particular relaxation properties, should be considered (see Section 6.2.3). The same applies to fatigue under alternating load (see Section 6.2.6).

For electrical and/or electronic applications, relaxation behavior of the connectors, the quality of the soldering joints and corrosion of the contacts are among others responsible for failures of the board circuit. As it is well known that a large part of the break-downs of the board circuit is to be blamed on malfunctions and failures of electric or electronic components, attention should be paid to the utilization of connector materials with good relaxation properties in critical areas and to good practice for the solderings as described in Section 7.2.1. Although copper materials show very good corrosion resistance, contacts in critical areas must be tin-plated or cladded with noble metals to ensure safe contacts also under severe corrosive conditions, e.g. by corrosive de-icing salt. Contact between vehicle manufacturers, suppliers and producers of semi-finished products is recommended as early as possible in the course of new developments.

Reliability of plain bearing materials depends significantly on the service life of the sliding materials. Manufacturers of plain bearings take account of increasing loads on plain bearings by more efficient materials or material combinations. Work is being done within plain bearing manufacturers' development departments on developing materials with considerably higher fatigue strength, lower wear rates and good corrosion resistance at higher temperatures. As soon as such higher demands are in sight, contact as soon as possible between vehicle and engine manufacturers and plain bearing suppliers is recommended – as has been practised up to now.

10.5
Repairability

The repairability of copper materials in automotive engineering is relevant only in a minimum of cases. The nature of copper materials and the kind of operational demands in automotive engineering normally lead to replacement of rundown, defect or worn copper material parts by new ones. In the design stage it should be kept in mind that the replacement should be possible with a minimum of expenditure.

Radiators and heat exchangers produced by the *Cu*proBraze® process, as they are brazed, could be repaired by soldering, which is not possible for aluminum-radiators. The economic feasibility of a repair would have to be verified. But in regions far from spare parts supply it might be helpful.

References

1 Kupfer – Vorkommen, Gewinnung, Eigenschaften, Verarbeitung, Verwendung, DKI brochure i.004, German Copper Institute, Düsseldorf, 1997.

2 Die Welt der Metalle; Kupfer – Geographie, Bergbau, Verhüttung, Handel, Metallgesellschaft AG, Frankfurt/Main, 1993.

3 Edelstein, D. L.; Copper. U. S. Geological Survey, Mineral Commodity Summaries, January 2004.

4 Arpaci, E.; Geschichte, Vorkommen und Gewinnung des Kupfers, Praxis der Naturwissenschaften – Chemie, 39(4) (1990), 2–9.

5 Fabian, H.; Kupfer, Bd. 15 in Ullmanns Encyclopaedia of Technical Chemistry, 4th edn., Verlag Chemie, Weinheim (1978), pp. 487–545.

6 Metallstatistik 2003. Wirtschafts-Vereinigung Metalle, 2004.

7 Natürlich Kupfer. Kupfer ökologisch gesehen. DKI brochure German Copper Institute, Düsseldorf, 1996.

8 Recycling von Kupferwerkstoffen. DKI-Informationsdruck i.027, German Copper Institute, Berlin, 1991.

9 Arpaci, E.; Vendura, T.; Recycling von Kupferwerkstoffen. VDI Report No. 917, 1992, pp. 595–616.

10 Arpaci, E.; Vendura, T.; Recycling von Kupferwerkstoffen – Klassische und neue Wiedergewinnungsverfahren, Metall, 47(4) (1993), 340–345.

11 Claassen, D., Kupfer bleibt bis 2005 knapp und teuer, Handelsblatt.com, 2005.

12 Krüger, J.; Reuter, M.; Winkler, P. et al., Sachbilanz einer Ökobilanz der Kupfererzeugung aus primären und sekundären Vorstoffen, sowie der Verarbeitung von Kupfer und Kupferlegierungen zu Halbzeug und ausgewählten Produkten, Metall 49 (1995) 4–6, Zugleich DKI reprint p. 198, German Copper Institute, Düsseldorf.

13 Zahlen und Daten der Entsorgungswirtschaft, Ch. 6: Recycling-Altautos. Bundesverband der Deutschen Entsorgungswirtschaft [Federal Waste Disposal Association], Sept. 2004.

14 Wieland-Kupferwerkstoffe (Herstellung, Eigenschaften und Verarbeitung). Wieland-Werke AG Metallwerke, Ulm, 1999.

15 Niedriglegierte Kupferwerkstoffe: Eigenschaften – Verarbeitung – Verwendung. DKI brochure i.008, German Copper Institute, Berlin/Düsseldorf.

16 Kupfer-Zink-Legierungen (Messing und Sondermessing). DKI brochure i.005, German Copper Institute, Berlin/Düsseldorf.

17 Kupfer-Zinn-Knetlegierungen (Zinnbronzen). DKI brochure i.015, German Copper Institute, Düsseldorf, 2004.

18 Kupfer-Nickel-Legierungen: Eigenschaften – Bearbeitung – Anwendung. DKI brochure i.014, German Copper Institute, Berlin/Düsseldorf.

19 Kupfer-Nickel-Zink-Legierungen (Neusilber). DKI brochure i.013, German Copper Institute, Berlin/Düsseldorf.

20 Kupfer-Aluminium-Legierungen: Eigenschaften – Herstellung, Verarbeitung – Anwendung. DKI brochure i.006, German Copper Institute, Berlin/Düsseldorf.

21 Kleinau, M.; Wenk, L.; Guss aus Kupfer und Kupferlegierungen: Technische Richtlinien, Gesamtverband Deutscher Metallgiessereien (GDM), Verein Deutscher Giessereifachleute [Associa-

Copper in the Automotive Industry. Hansjörg Lipowsky and Emin Arpaci

ISBN: 978-3-527-31769-1

tion of German Foundry Specialists] (VDG), German Copper Institute (DKI), Düsseldorf, March 1997.

22 Kupfer-Zinn-, Kupfer-Zinn-Zink- u. Kupfer-Blei-Zinn-Gusslegierungen (Guss-Zinnbronze, Rotguss und Guss-Zinn-Bleibronze). DKI brochure i.025, German Copper Institute, Berlin/Düsseldorf.

23 Bartz, W. J.; Gleitlagertechnik, Teil 1, 8.3 Kupferlegierungen Kontakt & et al. Studium, Bd. 49; expert verlag, 7031 Grafenau/Württemberg.

24 Köpf, H. et. al., Kupfer, 2. Auflage, Fachbuch,. German Copper Institute, Berlin/Düsseldorf, 1982.

25 Kupferwerkstoffe in der Elektrotechnik und Elektronik. DKI brochure i.010, German Copper Institute, Berlin/Düsseldorf.

26 Bänder und Drähte aus Kupferwerkstoffen für Bauelemente der Elektrotechnik und der Elektronik. DKI brochure i.020, German Copper Institute, Berlin/Düsseldorf.

27 Bänder, Bleche, Streifen aus Kupfer-Zink-Legierungen. DKI brochure i.022, German Copper Institute, Berlin/Düsseldorf.

28 Rohre aus Kupfer-Zink-Legierungen. DKI brochure i.021, German Copper Institute, Berlin/Düsseldorf.

29 Kupfer-Zink-Legierungen (Messing und Sondermessing). DKI specialist book (out of print), German Copper Institute, Berlin/Düsseldorf.

30 Brunhuber, E.; Guss aus Kupferlegierungen (Casting Copper-Base Alloys), from US English. Schiele & Schön GmbH, Berlin, 1986.

31 Kleinau, M.; Schwermetall-Schleuder- und Strangguss – technische und wirtschaftliche Möglichkeiten. DKI reprint p. 165, German Copper Institute, Berlin/Düsseldorf.

32 Material data sheets on various copper types low alloyed wrought copper alloys, wrought copper–zinc alloys, wrought copper–nickel alloys, wrought copper–tin alloys and copper casting alloys, German Copper Institute, Berlin/Düsseldorf.

33 Kleinau, M.; Kupfer und Kupferlegierungen für den Maschinenbau. Maschine und Werkzeug, 67 (1966), issue 28, pp. 9–16, Karl Ihl & Co, Coburg.

34 Arpaci, E.; Kupferlegierungen als Trägerwerkstoffe für Halbleiterbauelemente. Deutscher Ingenieurkalender, 1988, VDI Verlag, pp. 157–162.

35 Arpaci, E.; Bode A.; Kupferwerkstoffe – Eigenschaften und Anwendungen in der Elektrotechnik. Metall, 46(1) (1992), 22–31.

36 Arpaci, E.; Kupfer – Werkstoff der Zukunft; Trends der Technologie- und Marktentwicklung (Studie im Auftrag der Otto Brenner Stiftung) workbook No. 16, August 2000.

37 Löten von Kupfer und Kupferlegierungen. DKI brochure i.003 German Copper Institute, Berlin/Düsseldorf.

38 Stüer, H.; Biegeverhalten von Kupferlegierungsbändern; Recital: Kupferwerkstoffe für die Schwachstromtechnik, Wielandwerke AG, Ulm, 1984.

39 Böge, A.; Neuere Bandwerkstoffe aus Kupferlegierungen für Steckverbinder und Halbleiterträger. Bleche, Rohre, Profile, 39(12) (1992), 1042–1046.

40 Drefahl, K.; Kleinau, M.; Steinkamp, W.; Zeitstandeigenschaften und Bemessungskennwerte von Kupfer und Kupferlegierungen für den Apparatebau; DKI-Sonderdruck s. 178, German Copper Institute, Berlin/Düsseldorf.

41 Drefahl, K.; Kleinau, M.; Steinkamp, W.; Ergänzende Zeitstandversuche an den beiden Apparatewerkstoffen SF-CuF25 und CuZn20Al2. DKI reprint pp. 191, German Copper Institute, Berlin/Düsseldorf.

42 Thornton, C. H.; Harper, S.; Bowers, J. E.; INCRA Monograph XII The Metallurgy of Copper: A critical Survey of available high temperature mechanical property data for copper and copper alloys. International Copper Research Association, Inc., New York, 1983.

43 Copper Data Sheets, prepared by Conseil international pour le development du cuivre (CIDEC), Geneva, 1968.

44 Buss, B.; v. Meysenbug, C.-M.; Zeitstandversuche langer Dauer an Nichteisenmetallen, DECHEMA, Frankfurt, 1979

45 Lippmann, B.; Gegenwärtige und zukünftige Eigenschaftskriterien von Kupferbasiswerkstoffen und ihren Beschich-

tungen aus der Sicht eines Steckverbinderherstellers. Recital 3. Symposium Kupferwerkstoffe, 8./9. 10. 1992, Aachen. Proceedings of the German Copper Institute, Berlin/Düsseldorf.

46 Schleicher, K.; Dürrschnabel, W.; Bögel, A.; Kupferlegierungswerkstoffe als Basismaterialien für Bauteile in Elektrik und Elektronik, Recital 3. Symposium Kupferwerkstoffe 8./9. 10. 1992, Aachen. Proceedings of the German Copper Institute, Berlin/Düsseldorf.

47 Langer, H.; Relaxation von Kupferlegierungen – Einflüsse von Werkstoff und Zustand, Proceedings of the Kupferwerkstoffe für die Schwachstromtechnik, 1984, Wielandwerke AG, Ulm.

48 Helmenkamp, T.; Steinkamp, W.; Schleicher, K.; Spannungsrelaxationsmessungen an Bändern aus Kupferwerkstoffen. Metall, 43(11) (1989) pp. 1057–1061.

49 Murphy, M. C.; The Engineering Fatigue Properties of Wrought Copper. Fatigue of Materials and Structures, Vol. 4, Pergamon Press, UK, 1981, pp. 19–234.

50 Messner, O. H. C.; Werkstoffwahl und Werkstoffeinsatz, Pro Metall. III, 1974, Zurich.

51 Tiefziehen von Bändern und Blechen aus Kupfer und Messing. DKI reprint pp. 156, German Copper Institute, Berlin/Düsseldorf.

52 Biergans, W.; Fertigungstechniken, Eigenschaften und Verwendung von Stufenbändern. Recital 2. Symposium Kupferwerkstoffe, 20./21. 10. 1988, Berlin. Proceedings of the German Copper Institute, Berlin/Düsseldorf.

53 Richtwerte für die spanende Bearbeitung von Kupfer und Kupferlegierungen. DKI-Informationsdruck i.018, German Copper Institute, Berlin/Düsseldorf.

54 Kleinau, M.; Kupfer-Zink-Legierungen (Messing) für die Herstellung von Gesenkschmiedestücken (Warmpressteilen), Werkstofftechnische Anforderungen und metallkundliche Aspekte bei der Herstellung und Weiterverarbeitung von Kupfer-Zink-Legierungen. DKI reprint, p. 194, German Copper Institute, Berlin/Düsseldorf.

55 Greif, M.; Hochgeschwindigkeitsfräsen von Kupferlegierungen; Technologische Einflussgrössen und Randzoneneigenschaften. Dissertation, TH Darmstadt 1990, Carl Hanser, Munich, Vienna 1991.

56 König, W.; Spenrath, N.; Zerspantechnologie im Mikrometerbereich; Fertigung hochpräziser Reflexionsflächen. Industrie-Anz. 99/1988, pp. 33–36.

57 Hartel, R.; Spanende Bearbeitung von Metalloptikoberflächen. Industrie-Anzeiger 78/1988, pp. 30–31.

58 Förster, F.; Funkenerosives Drahtschneiden. Ein Fertigungsverfahren zur Herstellung von Prototypteilen, Kleinserien und Werkzeugen. Feinwerktech. Messtech. 88 (3) (1980) pp. 105–118.

59 Schweissen von Kupfer. DKI brochure i.012, German Copper Institute, Berlin/Düsseldorf.

60 Schweissen von Kupferlegierungen DKI brochure i.012 German Copper Institute, Berlin/Düsseldorf.

61 Kleben von Kupfer und Kupferlegierungen; DKI brochure i.007; German Copper Institute, Berlin/Düsseldorf.

62 Mechanische, chemische und elektrolytische Oberflächenvorbehandlung von Kupfer und Kupferlegierungen. DKI brochure i.023, German Copper Institute, Berlin/Düsseldorf.

63 Chemisches Farben von Kupfer und Kupferlegierungen. DKI brochure i.016, German Copper Institute, Berlin/Düsseldorf.

64 Chemische Färbungen von Kupfer und Kupferlegierungen. DKI manual, German Copper Institute, Berlin/Düsseldorf.

65 Emaillieren von Kupfer und Tombak; DKI reprint p. 163, German Copper Institute, Berlin/Düsseldorf.

66 Beschichten von Kupfer und Kupfer-Zink-Legierungen mit farblosen Transparentlacken. DKI brochure i.024, German Copper Institute, Berlin/Düsseldorf.

67 Simon, H.; Thoma, M.; Angewandte Oberflächentechnik für metallische Werkstoffe: Eignung – Verfahren – Prüfung. Carl Hanser, Munich, Vienna, 1985/89.

68 Schweissen von Kupfer und Kupferlegierungen; DKI manual, German Copper Institute, Berlin/Düsseldorf.

69 London Metal Exchange, daily stocks prices (*www.lme.com/dataprices*).

70 Pötzschke, M.; Grundlegende korrosionschemische Eigenschaften von Kupferwerkstoffen. DKI reprint p. 176, German Copper Institute, Berlin/Düsseldorf.

71 Dietrich, H.; Eigenschaften der nichtmagnetisierbaren NE-Metalle und ihre metallkundliche Deutung. Mitteilung aus dem Forschungsinstitut der Deutschen Edelstahlwerke AG, Krefeld; Metall, 20th edition IX. 96, H. 9, pp. 957–974.

72 Bögel, A.; Feind, J.; Gleitlagerwerkstoffe auf Kupferbasis für den Fahrzeugbau; Recital: 3. Symposium Kupferwerkstoffe, 8./9. 10. 92, Aachen; DKI-Tagungsband, German Copper Institute, Berlin/Düsseldorf.

73 Rübenach, F.; Selbstschmierende und wartungsfreie Gleitlager, J. Bartz (Hrsg.), Technische Akademie Esslingen, Esslingen, 1990.

74 Dörre, E.; Selbstschmierende und wartungsfreie Gleitlager, J. Bartz (Herausgeber), Technische Akademie Esslingen, 1990.

Appendix: Standards and Specifications

The following is a thematic summarization of few ISO standards and bodies of legislation and possibly the most important EN, ASTM and JIS standards concerning copper and copper alloys in the areas treated here. This is just a selection and the list is therefore not exhaustive. Information can be obtained via "Copper Key", www.copper-key.org, a database for comparison of International Standards. For further information and purchase of specifications please contact:

Deutsches Informationszentrum für Technische Regeln (DITR) im DIN
Burggrafenstrasse 6, D-10787 Berlin, Germany
Tel.: (+49)(0)190/00 2600 or (+49)(0)190/882600
Fax: (+49)(0)30/2628125; e-mail: Auskunft@din.de

British Standards Institution
389 Chiswick High Road
London
W4 4AL
England
Tel.: +44(0)2 089 969 001
e-mail: cservices@bsi-global.com
Website: www.bsi-global.com/index.xalter

American National Standards Institute – ANSI
1819 L Street, NW, 6th floor
Washington, DC 20036
USA
Tel.: 1 202 293 8020
Fax: 1 202 298 9287
e-mail: info@ansi.org
Website: www.ansi.org

Japanese Standards Association
4-1-24 Akasaka Minato-ku
Tokyo 107-8440, Japan
Tel.: +81-3-3583-8005
Website: www.jsa.or.jp/default_english.asp

Copper in the Automotive Industry. Hansjörg Lipowsky and Emin Arpaci

ISBN: 978-3-527-31769-1

A.1
EN-Standards

CEN/TS "Copper and copper alloys" was founded in 1988 for preparing and maintaining EN standards for blank shapes, semi-finished products and casting products of copper and copper alloys. Various product specifications are to be coordinated and rationalized in this context. The intention here was to set up a uniform, new and identifiable European classification system. Meanwhile, most of the work on these standards has been concluded. The work still to be done should be completed in the near future.

European equivalents (BS EN) of old British Standards relating to copper and copper alloys can be found in The Copper Development Association Publication No 120 "Copper and Copper Alloys – Compositions, Applications & Properties". Details of which along with other CDA Publications can be found at www.cda.org.uk.

A.1.1
Basic Standards

BS EN ISO 1302	Geometrical product specifications – Indication of surface texture in technical product documentation
EN 1173	Copper and copper alloys – Material condition and temper designation
EN 1412	Copper and copper alloys – European numbering system
EN 1655	Copper and copper alloys – Declarations of conformity
CR 12776	Copper and copper alloys – Provisions and procedures for the allocation of material numbers and registration of materials
CEN/TS 13388	Copper and copper alloys – Compendium of compositions and products (initial standard, September 2004)

A.1.2
Testing Standards

EN 10002-1	Tensile testing of metallic materials – Part 1: Method of test at ambient temperature
EN 10002 Part 5	Tensile testing of metallic materials – Part 1: Method of test at elevated temperature
EN 10204	Metallic products – Types of inspection documents
EN 12384	Copper and copper alloys – Determination of spring bending limit on strip

EN 13603	Copper and copper alloys – Test methods for assessing protective tin coatings on drawn round copper wire for electrical purposes
EN 14977	Copper and copper alloys – Locating tensile stress 5% ammonia test
EN ISO 196	Copper and wrought copper alloys – Locating residual stress – Mercury(I) nitrate test
EN ISO 2624	Copper and copper alloys – determination of average grain size
EN ISO 2626	Copper – hydrogen-embrittlement test
EN ISO 6509	Metal corrosion – Determining dezincification resistance of copper–zinc alloys
ISO 6957	Copper alloys – Ammonia test for stress corrosion resistance

A.1.3 Refinery Copper

EN 1976	Copper and copper alloys – Cast unwrought copper products
EN 1977	Copper and copper alloys – Copper drawing stock (wire rod)
EN 1978	Copper and copper alloys – copper cathodes

A.1.4 Flat Rolled Products

EN 1652	Copper and copper alloys – Plate, sheet, strip and circles for general use
EN 1653	Copper and copper alloys – Plate, sheet and circles for boilers, pressure vessels and hot water storage units
EN 1654	Copper and copper alloys – Strip for springs and connectors
EN 1758	Copper and copper alloys – Strip for lead frames
EN 13148	Copper and copper alloys – Hot-dip tinned strip
EN 14436	Copper and copper alloys – Electrolytically tin-plated strips

A.1.5 Tubes

EN 1057	Copper and copper alloys – Seamless round copper tubes for water and gas in sanitary- and heating applications
EN 12449	Copper and copper alloys – Seamless round tubes for general purposes
EN 12450	Copper and copper alloys – Seamless round copper capillary tubes
EN 12451	Copper and copper alloys – Seamless round tubes for heat exchangers
EN 12452	Copper and copper alloys – Rolled, finned, seamless, tubes for heat exchangers

EN 12735-1	Copper and copper alloys – Seamless round copper tubes for air conditioning and refrigeration – Part 1: Tubes for piping systems
EN 12735-1/A1	Copper and copper alloys – Seamless round copper tubes for air conditioning and refrigeration – Part 1: Tubes for piping systems
EN 12735-2	Copper and copper alloys – Seamless round copper tubes for air conditioning and refrigeration – Part 2: Tubes for equipment
EN 12735-2/A1	Copper and copper alloys – Seamless round copper tubes for air conditioning and refrigeration – Part 2: Tubes for equipment

A.1.6
Rods, Profiles and Wires

EN 12163	Copper and copper alloys – Rods for general purposes
EN 12164	Copper and copper alloys – Rods for free machining purposes
EN 12166	Copper and copper alloys – Wire for general purposes
EN 12167	Copper and copper alloys – Profiles and rectangular bars for general purposes
EN 12168	Copper and copper alloys – Hollow rods for free machining purposes

A.1.7
Forgings and Forging Stock

EN 12165	Copper and copper alloys – Wrought and unwrought forging stock
EN 12420	Copper and copper alloys – Forgings

A.1.8
Semi-finished Products for Electrical Engineering

EN 13599	Copper and copper alloys – Copper plate, sheet and strip for electrical purposes
EN 13600	Copper and copper alloys – Seamless copper tubes for electrical purposes
EN 13601	Copper and copper alloys – Copper rod, bar and wire for general electrical purposes
EN 13602	Copper and copper alloys – Drawn, round copper wire for the manufacture of electrical conductors

EN 13604 Copper and copper alloys – Products of high conductivity copper for electronic tubes, semiconductor devices and vacuum applications
EN 13605 Copper and copper alloys – Copper profiles and profiled wire for electrical purposes

A.1.9
Scrap, Master-alloys, Ingots and Castings

EN 1981 Copper and copper alloys – Master alloys
EN 1982 Copper and copper alloys – Ingots and castings
EN 1559-1 Founding – Technical conditions of delivery – Part 1: General
EN 12861 Copper and copper alloys – Scrap
ISO 8062 Castings – System of dimensional tolerances and machining allowances

A.1.10
Bearing Materials

ISO 4379 Plain bearings – Copper alloy bushings
ISO 4382-1 Plain bearings – Copper alloys – Part 1: Cast copper alloys for solid and multilayer thick-walled plain bearings
ISO 4382-2 Plain bearings – Copper alloys – Part 2: Wrought copper alloys for solid plain bearings
ISO 4383 Plain bearings – Multilayer materials for thin-walled plain bearings
ISO 5755 Sintered materials – Specifications

A.1.11
Soldering, Brazing and Welding

EN 1044 Brazing – Filler metals
EN 1045 Brazing – Fluxes for brazing – Classification and technical delivery conditions
EN 13347 Copper and copper alloys – Rod and wire for welding and braze welding
EN 14640 Welding additives – Solid wires and rods for fusion welding of copper and copper alloys
EN 29453 Soft solder alloys – Chemical compositions and forms
EN 29454-1 Soft soldering fluxes – Classification and requirements – Part 1: Classification, labelling and packaging
EN ISO 544 Welding consumables. Technical delivery conditions for welding filler metals. Type of product, dimensions, tolerances and marking

A.1.12
Other Technical Rules for Copper Materials

AD Data sheet W6/2	Materials for pressure vessels: Copper and wrought copper alloys, May 1988 edn. Carl Heymanns Verlag KG, Luxemburger Straße 449, Cologne. Beuth-Verlag GmbH, Burggrafenstraße 4–10, D-10787 Berlin
VdTÜV Material sheet 410	Installation tubes seamless drawn of SF-CuF37, June 1993 edn. Verband der Technischen Überwachungs-Vereine e.V., Essen. Printing and distribution: Verlag TÜV-Rheinland, PO Box 903060, D-51123 Cologne
Data sheet DVS 2903	Electrodes for resistance welding; Deutscher Verband für Schweisstechnik, Aachener Straße 172, D-40223 Düsseldorf, 10.1974
DG Worksheet (now DGO)	Influence of the basic material copper and its alloys on the result of galvanotechnical production. Deutsche Gesellschaft für Galvanotechnik, Düsseldorf, 03.1982

A.2
US-Standards (ASTM)

A.2.1
Basic Standards

ASTM B 224	Standard Classification of Coppers
ASTM B 275	Standard Practice for Codification of Certain Non-Ferrous Metals and Alloys, Cast and Wrought
ASTM B 601	Standard Classification for Temper Designations for Copper and Copper Alloys – Wrought and Cast
ASTM B 846	Standard Terminology for Copper and Copper Alloys
ASTM E 527	Standard Practice for Numbering Metals and Alloys (UNS)

A.2.2
Testing Standards

ASTM B 154	Standard Test Method for Mercurous Nitrate Test for Copper and Copper Alloys
ASTM B 279	Standard Test Method for Stiffness of Bare Soft Square and Rectangular Copper and Aluminum Wire for Magnet Wire Fabrication
ASTM B 577	Standard Test Methods for Detection of Cuprous Oxide (Hydrogen Embrittlement Susceptibility) in Copper

ASTM B 593	Standard Test Method for Bending Fatigue Testing for Copper-alloy Spring Materials
ASTM B 598	Standard Practice for Determining Offset Yield Strength in Tension for Copper Alloys
ASTM B 754	Standard Test Method for Measuring and Recording the Deviations from Flatness in Copper and Copper Alloy Strip
ASTM B 858	Standard Test Method for Ammonia Vapor Test for Determination of Susceptibility to Stress Corrosion Cracking in Copper Alloys
ASTM D 1688	Standard Test Methods for Copper in Water
ASTM D 1825	Standard Practice for Etching and Cleaning Copper-clad Electrical Insulating Materials and Thermosetting Laminate for Electrical Testing
ASTM D 1838	Standard Test Method for Copper Strip Corrosion by Liquefied Petroleum (LP) Gases
ASTM D 3482	Standard Test Method for Determining Electrolytic Corrosion of Copper by Adhesives
ASTM D 4048	Standard Test Method for Detection of Copper Corrosion from Lubricating Grease
ASTM E 8M	Standard Test Methods for Tension Testing of Metallic Materials [Metric]
ASTM E 53	Standard Test Methods for Determination of Copper in Unalloyed Copper by Gravimetry
ASTM E 62	Standard Test Methods for Chemical Analysis of Copper and Copper Alloys (Photometric Methods)
ASTM E 75* ANSI E 75	Standard Test Methods for Chemical Analysis of Copper–Nickel and Copper–Nickel–Zinc Alloys
ASTM E 106	Standard Test Methods for Chemical Analysis of Copper–Beryllium Alloys
ASTM E 112	Standard Test Methods for Determining Average Grain Size
ASTM E 118	Standard Test Methods for Chemical Analysis of Copper–Chromium Alloys
ASTM E 121	Standard Test Methods for Chemical Analysis of Copper–Tellurium Alloys
ASTM E 243	Standard Practice for Electromagnetic (Eddy Current) Examination of Copper and Copper Alloy Tubes
ASTM E 255	Standard Practice for Sampling Copper and Copper Alloys for the Determination of Chemical Composition
ASTM E 478	Standard Test Methods for Chemical Analysis of Copper Alloys
ASTM E 511	Standard Test Method for Measuring Heat Flux Using a Copper–Constantan Circular Foil, Heat-flux Gage
ASTM E 1606	Standard Practice for Electromagnetic (Eddy Current) Examination of Copper Redraw Rod for Electrical Purposes

ASTM G 37	Standard Practice for Use of Mattsson's solution of pH 7.2 to evaluate the Stress-corrosion Cracking Susceptibility of Copper–Zinc Alloys

A.2.3
Refinery Copper

ASTM B 5	Standard Specification for High Conductivity Tough-pitch Copper Refinery Shapes
ASTM B 115	Standard Specification for Electrolytic Copper Cathode
ASTM B 170	Standard Specification for Oxygen-Free Electrolytic Copper – Refinery Shapes
ASTM B 216	Standard Specification for Tough-pitch Fire-Refined Copper – Refinery Shapes
ASTM B 379	Standard Specification for Phosphorized Coppers – Refinery Shapes
ASTM E 1771	Standard Test Method for Determination of Copper in Anode and Blister Copper
ASTM E 1833	Standard Practice for Sampling of Blister Copper in Cast Form for Determination of Chemical Composition

A.2.4
Flat Rolled Products

ASTM B 19	Standard Specification for Cartridge Brass Sheet, Plate, Bar, and Disks
ASTM B 36/ B 36M	Standard Specification for Brass Plate, Sheet, Strip, and Rolled Bar
ASTM B 96/ B 96M	Standard Specification for Copper–Silicon Alloy Plate, Sheet, Strip and Rolled Bar for General Purposes and Pressure Vessels
ASTM B 100a	Standard Specification for Wrought Copper-alloy Bearing and Expansion Plates and Sheets for Bridge and Other Structural Use
ASTM B 101	Standard Specification for Lead-Coated Copper Sheet and Strip for Building Construction
ASTM B 103/ B 103M	Standard Specification for Phosphor Bronze Plate, Sheet, Strip, and Rolled Bar
ASTM B 171/ B 171M	Standard Specification for Copper-alloy Plate and Sheet for Pressure Vessels, Condensers, and Heat Exchangers
ASTM B 194	Standard Specification for Copper–Beryllium Alloy Plate, Sheet, Strip, and Rolled Bar
ASTM B 248	Standard Specification for General Requirements for Wrought Copper and Copper-alloy Plate, Sheet, Strip, and Rolled Bar

ASTM B 248M	Standard Specification for General Requirements for Wrought Copper and Copper-Alloy Plate, Sheet, Strip and Rolled Bar [Metric]
ASTM B 272	Standard Specification for Copper Flat Products with Finished (Rolled or Drawn) Edges (Flat Wire and Strip)
ASTM B 370	Standard Specification for Copper Sheet and Strip for Building Construction
ASTM B 422	Standard Specification for Copper–Aluminum–Silicon–Cobalt Alloy, Copper–Nickel–Silicon–Magnesium Alloy, Copper–Nickel–Silicon Alloy, Copper–Nickel–Aluminum–Magnesium Alloy, and Copper–Nickel–Tin Alloy Sheet and Strip
ASTM B 465	Standard Specification for Copper–Iron Alloy Plate, Sheet, Strip, and Rolled Bar
ASTM B 508	Standard Specification for Copper Alloy Strip for Flexible Metal Hose
ASTM B 534	Standard Specification for Copper–Cobalt–Beryllium Alloy and Copper–Nickel–Beryllium Alloy Plate, Sheet, Strip, and Rolled Bar
ASTM B 591	Standard Specification for Copper–Zinc–Tin and Copper–Zinc–Tin–Iron–Nickel Alloys Plate, Sheet, Strip and Rolled Bar
ASTM B 592	Standard Specification for Copper–Zinc–Aluminum–Cobalt Alloy, Copper–Zinc–Tin–Iron Alloy Plate, Sheet, Strip, and Rolled Bar
ASTM B 740	Standard Specification for Copper–Nickel–Tin Spinodal Alloy Strip
ASTM B 747	Standard Specification for Copper–Zirconium Alloy Sheet and Strip
ASTM B 768	Standard Specification for Copper–Cobalt–Beryllium Alloy and Copper–Nickel–Beryllium Alloy Strip and Sheet
ASTM B 820	Standard Test Method for Bend Test for Determining the Formability of Copper and Copper Alloy Strip
ASTM B 888	Standard Specification for Copper Alloy Strip for Use in the Manufacture of Electrical Connectors or Spring Contacts
ASTM B 936	Standard Specification for Copper–Chrome–Iron–Titanium Alloy Plate, Sheet, Strip and Rolled Bar

A.2.5 Tubes

ASTM B 42	Standard Specification for Seamless Copper Pipe, Standard Sizes
ASTM B 43	Standard Specification for Seamless Red Brass Pipe, Standard Sizes
ASTM B 68	Standard Specification for Seamless Copper Tube, Bright Annealed

ASTM B 68M	Standard Specification for Seamless Copper Tube, Bright Annealed [Metric]
ASTM B 75	Standard Specification for Seamless Copper Tube
ASTM B 75M	Standard Specification for Seamless Copper Tube [Metric]
ASTM B 88	Standard Specification for Seamless Copper Water Tube
ASTM B 88M	Standard Specification for Seamless Copper Water Tube [Metric]
ASTM B 111/ B 111M	Standard Specification for Copper and Copper-Alloy Seamless Condenser Tubes and Ferrule Stock
ASTM B 135	Standard Specification for Seamless Brass Tube
ASTM B 135M	Standard Specification for Seamless Brass Tube [Metric]
ASTM B 153	Standard Test Method for Expansion (Pin Test) of Copper and Copper-Alloy Pipe and Tubing
ASTM B 188	Standard Specification for Seamless Copper Bus Pipe and Tube
ASTM B 251	Standard Specification for General Requirements for Wrought Seamless Copper and Copper-Alloy Tube
ASTM B 251M	Standard Specification for General Requirements for Wrought Seamless Copper and Copper-Alloy Tube [Metric]
ASTM B 280	Standard Specification for Seamless Copper Tube for Air Conditioning and Refrigeration Field Service
ASTM B 302	Standard Specification for Threadless Copper Pipe, Standard Sizes
ASTM B 306	Standard Specification for Copper Drainage Tube
ASTM B 315	Standard Specification for Seamless Copper Alloy Pipe and Tube
ASTM B 359/ B 359M	Standard Specification for Copper and Copper-alloy Seamless Condenser and Heat Exchanger Tubes with Integral Fins
ASTM B 360	Standard Specification for Hard-drawn Copper Capillary Tube for Restrictor Applications
ASTM B 395/ B 395M	Standard Specification for U-Bend Seamless Copper and Copper Alloy Heat Exchanger and Condenser Tubes
ASTM B 428	Standard Test Method for Angle of Twist in Rectangular and Square Copper and Copper Alloy Tube
ASTM B 447	Standard Specification for Welded Copper Tube
ASTM B 466/ B 466M	Standard Specification for Seamless Copper–Nickel Pipe and Tube
ASTM B 467	Standard Specification for Welded Copper–Nickel Pipe
ASTM B 543	Standard Specification for Welded Copper and Copper-alloy Heat Exchanger Tube
ASTM B 569	Standard Specification for Brass Strip in Narrow Widths and Light Gage for Heat-Exchanger Tubing
ASTM B 587	Standard Specification for Welded Brass Tube
ASTM B 608	Standard Specification for Welded Copper-Alloy Pipe
ASTM B 640	Standard Specification for Welded Copper Tube for Air Conditioning and Refrigeration Service

ASTM B 643	Standard Specification for Copper–Beryllium Alloy Seamless Tube
ASTM B 687	Standard Specification for Brass, Copper, and Chromium-plated Pipe Nipples
ASTM B 698	Standard Classification for Seamless Copper and Copper Alloy Plumbing Pipe and Tube
ASTM B 706	Standard Specification for Seamless Copper Alloy (UNS No. C69100) Pipe and Tube
ASTM B 743	Standard Specification for Seamless Copper Tube in Coils
ASTM B 828	Standard Practice for Making Capillary Joints by Soldering of Copper and Copper Alloy Tube and Fittings
ASTM B 837	Standard Specification for Seamless Copper Tube for Natural Gas and Liquefied Petroleum (LP) Gas Fuel Distribution Systems
ASTM B 903	Standard Specification for Seamless Copper Heat Exchanger Tubes with Internal Enhancement
ASTM B 919	Standard Specification for Welded Copper Heat Exchanger Tubes with Internal Enhancement
ASTM B 937	Standard Specification for Copper-Beryllium Seamless Tube (UNS Nos. C17500 and C17510)

A.2.6 **Wires**

ASTM B 1	Standard Specification for Hard-Drawn Copper Wire
ASTM B 2	Standard Specification for Medium-Hard-Drawn Copper Wire
ASTM B 3	Standard Specification for Annealed Copper Wire
ASTM B 33	Standard Specification for Tinned Soft or Annealed Copper Wire for Electrical Purposes
ASTM B 47a	Standard Specification for Copper Trolley Wire
ASTM B 48	Standard Specification for Soft Rectangular and Square Bare Copper Wire for Electrical Conductors
ASTM B 99/ B 99M	Standard Specification for Copper–Silicon Alloy Wire for General Applications
ASTM B 105	Standard Specification for Hard-Drawn Copper Alloy Wires for Electric Conductors
ASTM B 134/ B 134M	Standard Specification for Brass Wire
ASTM B 159/ B 159M	Standard Specification for Phosphor Bronze Wire
ASTM B 189	Standard Specification for Lead-Coated and Lead-Alloy-Coated Soft Copper Wire for Electrical Purposes
ASTM B 197/ B 159M	Standard Specification for Copper–Beryllium Alloy Wire
ASTM B 206/	Standard Specification for Copper–Nickel–Zinc Alloy (Nickel

B 206M	Silver) Wire and Copper–Nickel Alloy Wire
ASTM B 246	Standard Specification for Tinned Hard-Drawn and Medium-hard-drawn Copper Wire for Electrical Purposes
ASTM B 250/ B 250M	Standard Specification for General Requirements for Wrought Copper Alloy Wire
ASTM B 272	Standard Specification for Copper Flat Products with Finished (Rolled or Drawn) Edges (Flat Wire and Strip)
ASTM B 298	Standard Specification for Silver-Coated Soft or Annealed Copper Wire
ASTM B 301/ B 301M	Standard Specification for Free-Cutting Copper Rod, Bar, Wire and Shapes
ASTM B 355	Standard Specification for Nickel-Coated Soft or Annealed Copper Wire
ASTM B 624	Standard Specification for High-Strength, High-Conductivity Copper-Alloy Wire for Electronic Application

A.2.7
Rods, Bars and Shapes

ASTM B 16/ B 16M	Standard Specification for Free-Cutting Brass Rod, Bar and Shapes for Use in Screw Machines
ASTM B 19	Standard Specification for Cartridge Brass Sheet, Plate, Bar, and Disks
ASTM B 21/ B 21M	Standard Specification for Naval Brass Rod, Bar, and Shapes
ASTM B 36/ B 36M	Standard Specification for Brass Plate, Sheet, Strip, and Rolled Bar
ASTM B 49	Standard Specification for Copper Rod Drawing Stock for Electrical Purposes
ASTM B 96/ B 96M	Standard Specification for Copper–Silicon Alloy Plate, Sheet, Strip and Rolled Bar for General Purposes and Pressure Vessels
ASTM B 98/ B 98M	Standard Specification for Copper–Silicon Alloy Rod, Bar and Shapes
ASTM B 103/ B 103M	Standard Specification for Phosphor Bronze Plate, Sheet, Strip, and Rolled Bar
ASTM B 121/ B 121M	Standard Specification for Leaded Brass Plate, Sheet, Strip and Rolled Bar
ASTM B 122/ B 122M	Standard Specification for Copper–Nickel–Tin Alloy, Copper–Nickel–Zinc Alloy (Nickel Silver), and Copper–Nickel Alloy Plate, Sheet, Strip and Rolled Bar
ASTM B 124/ B 124M	Standard Specification for Copper and Copper Alloy Forging Rod, Bar, and Shapes
ASTM B 138/ B 138M	Standard Specification for Manganese Bronze Rod, Bar, and Shapes

ASTM B 139/ B 139M	Standard Specification for Phosphor Bronze Rod, Bar, and Shapes
ASTM B 140/ B 140M	Standard Specification for Copper–Zinc–Lead (Red Brass or Hardware Bronze) Rod, Bar, and Shapes
ASTM B 150/ B 150M	Standard Specification for Aluminum Bronze Rod, Bar, and Shapes
ASTM B 151/ B 151M	Standard Specification for Copper–Nickel–Zinc Alloy (Nickel Silver) and Copper–Nickel Rod and Bar
ASTM B 152/ B 152M	Standard Specification for Copper Sheet, Strip, Plate, and Rolled Bar
ASTM B 169/ B 169M	Standard Specification for Aluminum Bronze Sheet, Strip, and Rolled Bar
ASTM B 187/ B 187M	Standard Specification for Copper, Bus Bar, Rod, and Shapes and General Purpose Rod, Bar, and Shapes
ASTM B 194	Standard Specification for Copper–Beryllium Alloy Plate, Sheet, Strip, and Rolled Bar
ASTM B 196/ B 196M	Standard Specification for Copper–Beryllium Alloy Rod and Bar
ASTM B 248	Standard Specification for General Requirements for Wrought Copper and Copper-Alloy Plate, Sheet, Strip, and Rolled Bar
ASTM B 248M	Standard Specification for General Requirements for Wrought Copper and Copper-Alloy Plate, Sheet, Strip and Rolled Bar [Metric]
ASTM B 249/ B 249M	Standard Specification for General Requirements for Wrought Copper and Copper-Alloy Rod, Bar, Shapes and Forgings
ASTM B 301/ B 301M	Standard Specification for Free-cutting Copper Rod, Bar, Wire and Shapes
ASTM B 371/ B 371M	Standard Specification for Copper–Zinc–Silicon Alloy Rod
ASTM B 411/ B 411M	Standard Specification for Copper–Nickel–Silicon Alloy Rod and Bar
ASTM B 441	Standard Specification for Copper–Cobalt–Beryllium and Copper–Nickel–Beryllium Rod and Bar (UNS Nos. C17500 and C17510)
ASTM B 453/ B 453M	Standard Specification for Copper–Zinc–Lead Alloy (Leaded-brass) Rod, Bar, and Shapes
ASTM B 465	Standard Specification for Copper–Iron Alloy Plate, Sheet, Strip, and Rolled Bar
ASTM B 534	Standard Specification for Copper–Cobalt–Beryllium Alloy and Copper–Nickel–Beryllium Alloy Plate, Sheet, Strip, and Rolled Bar
ASTM B 591	Standard Specification for Copper–Zinc–Tin and Copper–Zinc–Tin–Iron–Nickel Alloys Plate, Sheet, Strip and Rolled Bar

ASTM B 592	Standard Specification for Copper–Zinc–Aluminum–Cobalt Alloy, Copper–Zinc–Tin–Iron Alloy Plate, Sheet, Strip, and Rolled Bar
ASTM B 929	Standard Specification for Copper–Nickel–Tin in Spinodal Alloy Rod and Bar
ASTM B 936	Standard Specification for Copper–Chrome–Iron–Titanium Alloy Plate, Sheet, Strip and Rolled Bar

A.2.8
Forgings and Extruded Profiles

ASTM B 124/ B 124M	Standard Specification for Copper and Copper Alloy Forging Rod, Bar, and Shapes
ASTM B 249/ B 249M	Standard Specification for General Requirements for Wrought Copper and Copper-Alloy Rod, Bar, Shapes and Forgings
ASTM B 283	Standard Specification for Copper and Copper-alloy Die Forgings
ASTM B 455	Standard Specification for Copper–Zinc–Lead Alloy (Leaded-brass) Extruded Shapes
ASTM B 570	Standard Specification for Copper–Beryllium Alloy (UNS Nos. C17000 and C17200) Forgings and Extrusions
ASTM B 870	Standard Specification for Copper–Beryllium Alloy Forgings and Extrusions Alloys (UNS Nos. C17500 and C17510)
ASTM B 938	Standard Specification for Copper–Beryllium Alloy Forgings and Extrusions (UNS Nos. C17500 and C17510)

A.2.9
Products for Electrical Engineering

ASTM B 8	Standard Specification for Concentric-Lay-Stranded Copper Conductors. Hard, Medium Hard, or Soft
ASTM B 172a	Standard Specification for Rope-Lay-Stranded Copper Conductors Having Bunch-Stranded Members, for Electrical Conductors
ASTM B 174	Standard Specification for Bunch-Stranded Copper Conductors for Electrical Conductors
ASTM B 226	Standard Specification for Cored, Annular, Concentric-Lay-Stranded Copper Conductors
ASTM B 360	Standard Specification for Hard-Drawn Copper Capillary Tube for Restrictor Applications
ASTM B 372	Standard Specification for Seamless Copper and Copper-Alloy Rectangular Wave Guide Tube
ASTM B 470	Standard Specification for Bonded Copper Conductors for Use in Hookup Wires for Electronic Equipment

ASTM B 496	Standard Specification for Compact Round Concentric-Lay-Stranded Copper Conductors
ASTM B 628	Standard Specification for Silver–Copper Eutectic Electrical Contact Alloy
ASTM B 685	Standard Specification for Palladium–Copper Electrical Contact Material
ASTM B 694	Standard Specification for Copper, Copper-Alloy, Copper-Clad Bronze (CCB), Copper-Clad Stainless Steel (CCS), and Copper-Clad Alloy Steel (CAS) Sheet and Strip for Electrical Cable Shielding
ASTM B 702	Standard Specification for Copper–Tungsten Electrical Contact Material
ASTM B 715	Standard Specification for Sintered Copper Structural Parts for Electrical Conductivity Applications
ASTM B 738	Standard Specification for Fine-Wire Bunch-Stranded and Rope-Lay Bunch-Stranded Copper Conductors for Use as Electrical Conductors
ASTM B 784	Standard Specification for Modified Concentric-Lay-Stranded Copper Conductors for Use in Insulated Electrical Cables
ASTM B 787/ B 787M	Standard Specification for 19 Wire Combination Unilay-Stranded Copper Conductors for Subsequent Insulation
ASTM B 835	Standard Specification for Compact Round Stranded Copper Conductors Using Single Input Wire Construction
ASTM B 902a	Standard Specification for Compressed Round Stranded Copper Conductors, Hard, Medium-Hard, or Soft Using Single Input Wire Construction
ASTM F 68	Standard Specification for Oxygen-free Copper in Wrought Forms for Electron Devices
ASTM F 96	Standard Specification for Electronic Grade Alloys of Copper and Nickel in Wrought Forms

A.2.10
Master Alloys, Ingots and Castings

ASTM B 22	Standard Specification for Bronze Castings for Bridges and Turntables
ASTM B 30	Standard Specification for Copper Alloys in Ingot Form
ASTM B 61	Standard Specification for Steam or Valve Bronze Castings
ASTM B 62	Standard Specification for Composition Bronze or Ounce Metal Castings
ASTM B 66	Standard Specification for Bronze Castings for Steam Locomotive Wearing Parts
ASTM B 148	Standard Specification for Aluminum Bronze Sand Castings
ASTM B 176	Standard Specification for Copper-Alloy Die Castings

ASTM B 208	Standard Practice for Preparing Tension Test Specimens for Copper Alloy Sand, Permanent Mold, Centrifugal, and Continuous Castings
ASTM B 271	Standard Specification for Copper-Base Alloy Centrifugal Castings
ASTM B 369	Standard Specification for Copper–Nickel Alloy Castings
ASTM B 427	Standard Specification for Gear Bronze Alloy Castings
ASTM B 505/ B 505M	Standard Specification for Copper-Base Alloy Continuous Castings
ASTM B 584	Standard Specification for Copper Alloy Sand Castings for General Applications
ASTM B 763	Standard Specification for Copper Alloy Sand Castings for Valve Applications
ASTM B 770	Standard Specification for Copper–Beryllium Alloy Sand Castings for General Applications
ASTM B 806	Standard Specification for Copper Alloy Permanent Mold Castings for General Applications
ASTM B 824	Standard Specification for General Requirements for Copper Alloy Castings
ASTM E 272	Standard Reference Radiographs for High-Strength Copper-Base and Nickel–Copper Alloy Castings
ASTM E 310	Standard Reference Radiographs for Tin Bronze Castings

A.2.11 Bearing Materials

ASTM B 436/ B 436M	Standard Specification for Sintered Bronze Bearings (Oil-Impregnated)
ASTM B 612	Standard Specification for Iron Bronze Sintered Bearings (Oil-Impregnated)

A.2.12 Soldering, Brazing and Welding

ASTM B 813	Standard Specification for Liquid and Paste Fluxes for Soldering of Copper and Copper Alloy Tube

A.2.13 Coatings

ASTM B 281	Standard Practice for Preparation of Copper and Copper-Base Alloys for Electroplating and Conversion Coatings
ASTM B 734	Standard Specification for Electrodeposited Copper for Engineering Uses
ASTM B 904	Standard Specification for Autocatalytic Nickel over Autocatalytic Copper for Electromagnetic Interference Shielding

A.3 Japanese Standards (JIS)

A.3.1 Basic Standards

JIS C 3001	Resistance of copper materials for electrical purposes
JIS H 0500	Glossary of terms used in wrought copper and copper alloys
JIS H 2109	Classification standard of copper and copper alloy scraps
JIS H 2501	Phosphor copper metal
JIS K 0010	Reference materials – Standard solution – Copper

A.3.2 Testing Standards

JIS C 3002	Testing methods of electrical copper and aluminum wires
JIS C 3003	Methods of test for enamelled wires
JIS C 3006	Methods of test for fiber or paper insulated wires
JIS H 0501	Methods for estimating average grain size of wrought copper and copper alloys
JIS H 0502	Method of eddy current testing for copper and copper alloy pipes and tubes
JIS H 0530	Measurement methods of polarization resistance of copper alloy tubes for condenser
JIS H 1064	Method for determination of tellurium in copper
JIS H 1065	Method for determination of selenium in copper
JIS H 1069	Methods for determination of cadmium in copper and copper alloys
JIS H 1101	Methods for chemical analysis of electrolytic cathode copper
JIS H 1552	Methods of chemical analysis for phosphor copper ingots
JIS Z 3900	Methods for sampling of precious brazing filler metals
JIS Z 3903	Methods for chemical analysis of copper phosphorus brazing filler metals

A.3.3 Refinery Copper

JIS H 2121	Electrolytic cathode copper

A.3.4 Flat Rolled Products

JIS C 2521	Copper nickel alloy wires, rolled wires, ribbons and sheets for electrical resistance

JIS C 2522	Copper–manganese alloy wires, bars and sheets for electrical resistance
JIS C 6515	Copper foil for printed wiring boards
JIS H 3100	Copper and copper alloy sheets, plates and strips
JIS H 3110	Phosphor bronze and nickel silver sheets, plates and strips
JIS H 3130	Copper–beryllium alloy, copper–titanium alloy, phosphor bronze and nickel silver sheets, plates and strips for springs
JIS H 3510	Oxygen-free copper sheet, plate, strip, seamless pipe and tube, rod, bar and wire for electron devices

A.3.5 Tubes

JIS B 2240	General rules for copper alloy pipe flanges
JIS H 3300	Copper and copper alloy seamless pipes and tubes
JIS H 3320	Copper and copper alloy welded pipes and tubes
JIS H 3401	Pipe fittings of copper and copper alloys
JIS H 3510	Oxygen free copper sheet, plate, strip, seamless pipe and tube, rod, bar and wire for electron devices

A.3.6 Wires

JIS C 2521	Copper–nickel alloy wires, rolled wires, ribbons and sheets for electrical resistance
JIS C 2522	Copper–manganese alloy wires, bars and sheets for electrical resistance
JIS C 2523	Oxidized copper–nickel alloy wires for electrical resistance use
JIS C 2528	Silk and polyester fiber covered wires for electrical resistance use
JIS C 2529	Enamelled wires and oleo-resinous enamelled silk covered wires for electrical resistance use
JIS C 2806	Non-insulated crimp-type sleeves for copper conductors
JIS C 3101	Hard-drawn copper wires for electrical purposes
JIS C 3102	Annealed copper wires for electrical purposes
JIS C 3103	Annealed copper wires for winding wires
JIS C 3104	Rectangular copper wires for electrical purposes
JIS C 3105	Hard-drawn copper stranded conductors
JIS C 3106	Copper wire rods for electrical purposes
JIS C 3152	Tin coated annealed copper wires
JIS C 3204	Fiber or paper-covered copper winding wires
JIS E 2101	Hard-drawn grooved trolley wires
JIS E 2102	Hard-drawn copper round trolley wires
JIS H 3260	Copper and copper alloy wires
JIS H 3270	Copper beryllium alloy, Phosphor bronze and nickel silver rods, bars and wires

JIS H 3510	Oxygen-free copper sheet, plate, strip, seamless pipe and tube, rod, bar and wire for electron devices

A.3.7 Rods, Bars and Shapes

JIS H 2123	Copper billets and cakes
JIS C 2522	Copper–manganese alloy wires, bars and sheets for electrical resistance
JIS H 3250	Copper and copper alloy rods and bars
JIS H 3270	Copper–beryllium alloy, Phosphor bronze and nickel silver rods, bars and wires
JIS H 3510	Oxygen-free copper sheet, plate, strip, seamless pipe and tube, rod, bar and wire for electron devices

A.3.8 Master-alloys, Ingots and Castings

JIS H 2202	Copper alloy ingots for castings
JIS H 5120	Copper and copper alloy castings
JIS H 5121	Copper alloy continuous castings

A.3.9 Soldering, Brazing and Welding

JIS Z 3202	Copper and copper alloy gas welding rods
JIS Z 3231	Copper and copper alloy covered electrodes
JIS Z 3234	Copper alloys for resistance welding electrode
JIS Z 3262	Copper and copper alloy brazing filler metals
JIS Z 3264	Copper phosphorus brazing filler metals
JIS Z 3341	Copper and copper alloy rods and solid wires for inert gas shielded arc welding
JIS Z 3621	Recommended practice for brazing

A.3.10 Products for Electrical Engineering

JIS C 2805	Crimp-type terminal lugs for copper conductors
JIS H 3140	Copper bus bars
JIS H 3510	Oxygen free copper sheet, plate, strip, seamless pipe and tube, rod, bar and wire for electron devices

A.3.11 Coatings

JIS H 8624	Electroplated coatings Tin-lead alloys
JIS H 8646	Electroless copper platings

Subject Index

Copper in the Automotive Industry. Hansjörg Lipowsky and Emin Arpaci

ISBN: 978-3-527-31769-1

d

y

z

Related Titles

Streitberger, H.-J., Kreis, W. (Eds.)

Automotive Paints and Coatings

2006
ISBN 3-527-30971-3

Hesse, J., Gardner, J.W., Göpel, W. (Series Editors)
Marek, J., Trah, H.-P., Suzuki, Y., Yokomori, I. (Volume Editors)

Sensors for Automotive Technology
Sensors Applications Vol. 4

2003
ISBN 3-527-29553-4

Bode, H. (Ed.)

Materials Aspects in Automotive Catalytic Converters

2002
ISBN 3-527-29773-1

Murarka, S.P., Verner, I.V., Gutmann, R.J.

Copper-Fundamental Mechanisms for Microelectronic Applications

2000
ISBN 0-471-25256-5

Poole, C.P.

Copper Oxide Superconductors

1988
ISBN 0-471-62342-3